SANTA ANA PUBLIC LIBRARY
D0450355

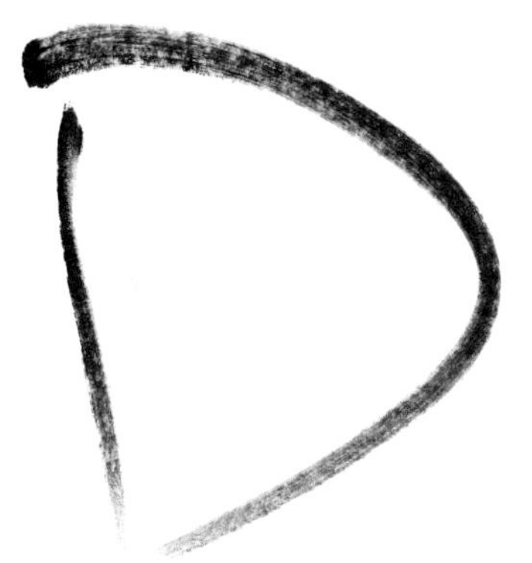

HOT
MESS

Matt Winning

HOT MESS

First published in 2021 by
HEADLINE PUBLISHING GROUP

1

Cataloguing in Publication Data is available from the British Library

Hardback ISBN 978 1 4722 7668 1
Trade paperback ISBN 978 1 4722 7669 8

Typeset in Monotype Sabon by CC Book Production
Printed and bound in Great Britain by Clays Ltd, Elcograf S.p.A.

Headline's policy is to use papers that are natural, renewable and recyclable products and made from wood grown in well-managed forests and other controlled sources. The logging and manufacturing processes are expected to conform to the environmental regulations of the country of origin.

HEADLINE PUBLISHING GROUP
An Hachette UK Company
Carmelite House
50 Victoria Embankment
London EC4Y 0DZ

www.headline.co.uk
www.hachette.co.uk

For BB – this book is my small effort to make your world
a better place. And it is your world now.
And WJ – I hope I make you proud the way you
make me proud every single day.

And to whomever invented Ben & Jerry's
Peanut Butter & Cookies vegan ice cream.
This book would not have been possible without you.

Contents

PART 2: CAN WE CHANGE?

PART 3: WILL WE CHANGE?

1

Nine months before

I feel calm, relaxed and at ease

'We're having a baby,' said my wife as she returned from the bathroom.

I'd always expected such news to be a defining moment in my life. For that instant to change everything, forever. But it didn't. Not yet. All that had changed was that my wife was standing in the middle of a room, sobbing tears of joy while clutching an object covered in urine. I was in the middle of eating my second Terry's Chocolate Orange of the day. And all I could really think about was that it was pretty odd that this was all happening in a stranger's home. You see, it was during that weird time between Christmas and New Year. Norwegians call it Romjul, and the eighteenth-century Scots poet Robert Fergusson aptly called it 'The Daft Days'. You know, when you eat your own weight in chocolate, barely see sunlight and tell friends you'll meet up with them, only to cancel at the last minute so you can stay in and watch *Raiders of the Lost Ark* for the umpteenth time. And due to some renovations at my parents' house, we were in an Airbnb flat in a part of Glasgow to which I would never normally venture. It was beside a high street dedicated to charity shops, next to a local pub that looked closed to the outside world, surrounded by impenetrable shutters, where the only sign of life was a Union Jack flying limply.

So, I guess it was just all a bit strange.

Don't get me wrong: this was wonderful news. Even since I was a teenager, the one thing I had been certain of in life is that one day I wanted to have a child. That, and I was certain that people who refer to guacamole as 'guac' should be put on a list. So, two things. I'm just not the most decisive person in the world. I don't mean that I'm one of those irritating people who cannot decide what to order in a restaurant, like my old schoolfriend Ian, who, without fail, waits for the waiter to arrive before infuriatingly going, 'Oh, do we need to choose now? You'd better ask the others first . . . Hmmm, I can't decide, can you tell me about the soup? Yes, but the melon does sound good,' as if he's never been in a bloody restaurant before.

I am incredibly indecisive when it comes to big life stuff, though. I can end up analysing too much. For instance, I wasn't sure what subject to do at university, so I started a joint degree. Thirty-two other people also started that specific joint degree. I was the only one to finish it. Not because I am great, but because nobody else was mad enough to continue with it, and instead they all rightly chose a career path. It's something I very much get from my father, a man who is a neurotic overthinker and pays inordinate attention to detail. I like to describe him as the sort of man who would open a party popper directly into the bin. I hope I don't pass this trait on.

I'd always wanted to become a dad. Having a wee person

I could direct my attention and affection towards just seemed like what life was all about. I mean, thinking about yourself is wonderful and all, but it also seems, well, selfish. I think I looked up to my dad, and to his dad, and aspired to be like them. Creating a family just felt like it would be the most rewarding journey – and why would you not want more people in your life that you love unconditionally?

Something had made me pause, though. I'd just spent the last decade researching the frightening reality of climate change. You see, I spend most of my waking days sitting at a desk, running computer models about what the hypothetical temperature of the planet will be in the year 2100 under varying energy system futures. It is an odd job. I mean, it is somewhat detached from the climate front line. Compared to, say, a scientist who works on drilling ice cores in Antarctica, it can seem like I just look at numbers on a screen. However, all these numbers have started turning into physical manifestations over the last couple of years: clear effects of climate change, happening all around the world. And this was increasingly allowing doubt to creep into my mind.

Let me be clear, as my wife stood there clutching the positive pregnancy test, I was still 100 per cent sure that I *wanted* to have a child. I had just been watching *The Mandalorian*, and Baby Yoda's adventures had convinced me that looking after a child was an exciting adventure. The doubt was around whether I *should* have a child, now, at this time in history (gestures around at, well, everything). Should we bring a child into the world? Was that a good or right thing to do – for others, for the child, for us? And, more importantly, should I have a third Terry's Chocolate Orange?

Apparently, a lot of other people feel this way too. The

stark reality has led to a new term used to describe when this worry becomes debilitating: 'eco-anxiety', which came close to winning the crown for the Oxford English Dictionary's Word of the Year 2019, only losing out to 'climate emergency'.[1] The American Psychological Association defined eco-anxiety as 'a chronic fear of environmental doom'.[2] Now, I don't think I have this, but maybe I am lucky, because I'm already doing some of the best things you can do to address such anxiety. I'm talking about it; I'm doing something about it; I'm taking action. I've pretty much structured my entire professional life towards addressing the issue. But worries can still creep in. I'm beginning to feel like perhaps there are two Matt Winnings battling it out inside my brain now: the rational level-headed academic and the emotional chocolate-binging father-to-be.

Groups are taking this anxiety and using it to push for real government action. This can be seen in the School Strike for Climate protests, started by the Swedish teenager Greta Thunberg, who has done as much for climate change as she has for the resurrection of the yellow rain mac industry. And groups have even popped up specifically around the idea of having – or not having – kids in the age of climate breakdown, such as BirthStrike and Conceivable Future. They are organisations led by women and are dedicated to grappling with reproductive choices during a climate crisis. Many believe that until we are on a clear path to a safer future, then having children is too risky.[3] The purpose of these groups is to provide a place for others who wish to share their stories and to highlight the climate issues to a wider audience. The other year, even Prince Harry stated that he and Meghan Markle will only have two children for the sake of the climate. I tried to imagine what the carbon footprint of a royal baby might be, what with all that flying across the world and State dinners. Although I don't know what lizards eat.

All these thoughts and concerns had been swimming around in the deep end of my brain for some time, but I tried to do what my west-of-Scotland upbringing had taught me: bottle up my emotions, then tie a heavy brick to that bottle and throw it into a loch of despair, letting it sink to the bottom along with all other regrets in my life: my relationship with my father; playing it too cool with a girl I liked; and the time I was ridiculed at primary school for thinking the Will Smith song 'Gettin' Jiggy Wit It' was called 'You Get a Chicken With It'. While having a baby had been hypothetical, I hadn't engaged with the idea too much, and all was fine. But now, like climate change, it was very much real and happening. And, like climate change, I needed to start preparing for it.

My wife and I hugged tightly. We were delighted and excited and scared and happy. Being in an Airbnb in Govan can do that. We were, both literally and figuratively, in the unknown. And we were entirely unprepared for what was to come. It had begun.

Oh, and there is one more thing I am sure about. That man-made global warming is happening, and that without immediate action it will irreversibly alter the planet for millennia to come, causing unnecessary suffering and pushing ecosystems to the brink of collapse, stretching our human capacity to respond to disaster into uncharted territory. Three things. I'm sure of three things.

2
Intro

'This is not comedy,' a bald audience member with a goatee mouthed to the woman sitting next to him.

I had been on stage for thirty minutes already, and was currently showing a PowerPoint slide of a carbon tax supply-and-demand graph, while talking about how Captain Planet clearly knew nothing about climate economics, because the optimal amount to 'take pollution down to' may not be precisely zero. I continued to attack the early-nineties environmental cartoon hero for the fact he hadn't specified whether he had meant 'net-zero', asking if he therefore didn't believe it was possible to scale up and deploy necessary levels of carbon capture and storage technologies.

To be fair, the audience member may have had a point.

We were at a comedy night, for God's sake, and these people were on a midweek night out in the back of a pub in Croydon. The capital of comedy. I don't think people realise how much comedians are multitasking while we're performing on stage. While we're talking, we are also thinking about what we're about to say next, and simultaneously looking around the room to make judgements about whether people are or aren't enjoying it, in order to change what we might do. In this situation, I could pretty much follow that this couple were not enjoying what was ostensibly a Scottish man in his early thirties giving a lecture about climate change – but this time there was no plan B. I had decided to commit to this concept for an hour, nevertheless.

I'd started performing comedy back in 2009. The previous year, I had started a PhD programme on climate policy, involuntarily split up with my long-term girlfriend, and begrudgingly moved back in with my parents. I needed to get out of the house. As often as possible. And to get a break away from my PhD. So, I started doing stand-up comedy.

Now, here I was, in February 2017, dying on my arse on stage in south London. I mean, some people were laughing at the sheer audacity and absurdity of it, but that couldn't sustain a full show. It was a night of early previews for the Edinburgh Fringe, where we were to try new material. I was to be onstage for an hour, then there'd be a break, and then another act. When the second act eventually came on after pints had been refreshed, he told the audience he was here to give a lecture about the ozone layer – and that got by far the biggest laugh of the night.

I had decided this year would be my last throw of the dice. I was in my early thirties now and, unlike most comedians with day jobs (usually in Oddbins), I actually enjoyed mine, and found meaning in my role as a climate researcher. I had intentionally not talked about climate onstage up until now – generally, the topic is a bit of a buzzkill. But I'd run out of ideas. People always say you should talk about what you know. So, after almost a decade, I decided that I was going to perform a comedy lecture about climate change for twenty-five days straight at the world's largest arts festival, come hell or high water (both of which are apt descriptions of the climate struggle), and it was clear it was going to be a challenge. Maybe I was in over my head. I'd had this idea of trying to make climate change funny. And it was clear from tonight's show that I didn't yet know how to do that.

Who am I?

Let me tell you a little bit about myself. My name is genuinely Dr Winning. I know, it sounds made up and probably what Charlie Sheen's drug dealer is listed under on his mobile phone, but sadly I am nothing so glamorous: I have a PhD in climate change policy. I always thought that having a PhD would change my life: that it'd be highly impressive to people; that once I became a doctor, everyone would respect me; that I'd be opening a door to a world of intellect and sophistication. But I now realise my error. The only way my life has altered is that family members now have a stick to hit me with when I do something idiotic. Which is often. I constantly lose my phone, and my wife says, 'How can you have lost it again? You have a PhD,' and I reply, 'But I don't have a PhD in not losing my phone,' and we both laugh that laugh we all do when change is futile.* Nobody respects you for having a PhD. I was once called a 'boffin' in a tabloid newspaper, a word that is intended to demean anyone who dares know anything about something.

I like to tell people that I am the sort of doctor that, if you have a heart attack on a flight and somebody asks 'Is there a doctor on board?', then I will rush to your side, but only to berate you for flying. I work at one of the world's leading research institutes on sustainability. Or, as my dad likes to tell people, 'Matthew is thirty-six and still at university.' My job title is technically 'environmental economist', which I've always found to be a bit of an oxymoron. It's a bit like being a human rights lawyer – you may be helping people, but ultimately you are *still* a lawyer – or it's an oxymoronic let-down, like being offered a cocktail . . . sausage. I thought it would mean I was a *good* economist, you know, one of the good guys. In practice, it turns out most normal people just hear the 'environmental' part and assume I'm some sort of a hippy. I wouldn't even really consider myself an environmentalist – for instance, I don't think burning incense is a substitute for having a personality. Meanwhile, most environmentalists just hear the 'economist' part and therefore assume I am the devil incarnate. But hey ho!

---

* NOTE: I want to make it clear from the start that the comedy in this book is absolutely not 'my wife'. We get on very well. I was about to say we get on 'like a house on fire', but I now realise how weird a phrase that is, and I have no idea what it really means or where it comes from. There goes thirty minutes of writing time, because now I have to google this (I say google, but I actually use the search engine Ecosia, as they plant trees when you use it, and this helps avert climate change).

Why have I written this book?

Because things are getting worse. Much worse. We need to talk about it. To me, climate change is a bit like Michael Jackson: we've known about the issues since the 1980s and collectively hoped they would just go away . . .

I believe climate change is the most wicked problem humans could possibly face. And by wicked, I don't mean like the musical. Although, I suppose, her face *is* green. What I mean by a 'wicked problem', is that if I were an evil mastermind intent on destruction and sat down to invent a way to destroy humanity from scratch, what I came up with would probably look a lot like climate change. It is invisible, so we don't really notice it. Yet it is pervasive. It is complex, so we get bored of it easily. It happens slowly, so we don't notice the danger. As dangers go, it's damn near the opposite of the kind we humans are good at detecting.

But we are also living through a time of unparalleled innovation and prosperity. Never before have we had a better shot at solving and reshaping our global society towards a cleaner path. Climate change is happening at both the worst and the best time possible.

What is the structure of the book, I hear you ask? Good question.

Because I am an academic, I like to tell you what I'm going to tell you, then I spend ages telling you it, and then at the end I tell you that I told you what I was going to tell you. I call it a 'research summary', writers call it a 'contents page', and young people call it 'spoilers'.

The overall structure of the book is split into the following three questions:

- Should We Change?
- Can We Change?
- Will We Change?

And these just happen to be the same three questions you will be asked at any job interview for a bureau de change.* In fact, I always find it much more uplifting to say the 'change' in 'climate change' in a French accent (pronounced '*shawn-ge*'). You've got to get your kicks somewhere when you are working on such a bleak topic. The first part, 'Should We Shawnge?', looks at the science and its impacts, to understand what is happening and so we are all on the same recycled page. In the second part, 'Can We Shawnge?', we delve into the solutions and what has to happen to avert the worst. In the third and final section, 'Will We Shawnge?', we explore why climate action is more complicated, why nothing seems to have been done and how this needs to be addressed.

I top and tail the book with the two questions I am asked most by friends and members of the public: 'Are we screwed?' and 'What can I do about it?' The first is mostly people just wanting to be reassured, and the second is a good question as the information available is often unclear and confusing. Suggestions are often lots of small things, like changing your light bulbs, recycling more, travelling back in time to kill yourself before you were even born, blah blah. But do they work?

* These three headings are borrowed from Al Gore's 2016 TED Talk 'The case for optimism on climate change'. Hope Alan (or Alistair or Albert?) doesn't mind? I found it a really helpful way to think about the issues, and I tried to find alternatives of my own, but sometimes it is better to admit someone else did it best and simply plagiarise. Plus, the bureau de change joke is so good I've decided to base the entire structure of the book around it.

Then, at the end of the book, there will be some sort of a summary. And I know what you are thinking: that I'm about to make a joke about how, in the future, thanks to climate change, every day will be summery. But I'm not going to make that joke, as it doesn't work written down. 'Summary' and 'summery' are spelled differently. Also, it's considerably more complicated than that. So there.

PART 1
SHOULD WE CHANGE?

3
Are we screwed?

It feels like we're screwed, doesn't it? That's the general 'vibe' we all get from climate change. The constant barrage of endless scary news stories and reports about how we are all doomed . . . but are we?

We're not good at knowing when to stop. Humans, that is. Whether it's having too many drinks celebrating the news that you're going to have your first child, or having too many drinks celebrating that you've finished writing a book. There's never too much of a good thing. Until there is. We are good at *doing* things. We're not great at *stopping* doing things. If we start doing this, and this is good, then this is who I am: I am a person that does this. And we continue in that vein. Collectively, we are even worse, as reinforcing mechanisms kick in. We become a society that does this. Once the ball gets rolling, it is hard to slow it down. Things have got out of control, haven't they? And now the hangover is here. The trick is finding new, better things to do.

People who ask me 'Are we screwed?' are probably hoping for good news, so they can stop worrying about it. Alternatively, they are quietly sadomasochists: the kind of people who 'love a good cry' or secretly believe they'd thrive in a post-apocalyptic wasteland. Only the latter tend to get what they want, as once you go down the rabbit hole* of what climate change entails, it is like innocently delving into your parents' bedroom drawers or joining social media, i.e. mostly horrific. And climate change is proving even more difficult for humanity to solve than deciding what colour a packet of salt and vinegar crisps should be.

It depends on who you are, but as a general rule, it's going to get tougher to exist as a human being. If you've bought this book, chances are you might be one of the people least directly affected by climate change. But while you might secretly believe you'll be able to hide out in a Waterstones café until it blows over, like it or not, every single person on earth is going to be affected in one way or another. Whether it is food shortages, more heatwaves, increased flooding, more political unrest, or simply the annual ski season being reduced to three days,† every society on earth is going to be pushed outside of its comfort zone. After all, our towns, our houses, our workplaces, our *lives* are specifically designed to be able to cope with 'normal' conditions, in a relatively stable climate, but what is 'normal' is changing – and rapidly. When heatwaves and floods that previously only occurred once every

---

* Two thirds of rabbit species are threatened by climate change, but while I'm keen to future-proof this book, 'down the unspecified remaining animal hole' doesn't quite have the same ring to it.

† With no respect for school half-terms.

hundred years start to happen every five years, our lives will have to change as well.

This obviously raises a lot of questions, like: How quickly can humanity adapt to these unpredictable and changing conditions? How difficult and costly will these changes be? And wouldn't it be better if we could just find a way of preventing these things from happening in the first place?

As someone who isn't 100 per cent confident about making plans for the weekend more than two days in advance, these questions can seem huge and scary, and it's quite literally my job to answer them. But I'm not the only one who has to answer them. We all do, and everyone on the planet – almost eight billion of us – has a different idea about what an acceptable level of change looks like, and what they should do about it. Put bluntly, humanity is at a crossroads. We're playing the world's shittiest *Choose Your Own Adventure*, and we can't decide if we should turn to page six and dump our old boyfriend, or page twenty-six, to meet a new boyfriend.*

You might not personally feel like you have a huge amount of say in what humanity as a whole decides to do about climate change, but the decisions we make now, and over the next decade, will determine the state of our descendants' lives on this planet for thousands of years to come. No biggie.

For me, the clearest way to think about our choices when it comes to climate change comes from John Holdren, who was Barack Obama's scientific advisor. He concluded that we have three basic choices: (1) adaptation, (2) mitigation, or (3) suffering. That is: (1) we change how we exist to live with

---

* I've only read one *Choose Your Own Adventure* book, and I think it was pretty specifically geared towards teenage girls.

it, (2) we do what we can to prevent it from happening, or (3) we do nothing, and suffer the impact. It is already inevitable that humanity will experience some mixture of all of these three outcomes, but we can still make choices that affect the proportions. The more action we take now, the less suffering will be felt, and the less we'll need to adapt to this new world.

How long have we got to make these decisions? In 2018, we were told that we have twelve years left to save the world, but let's break that number down a bit.

First of all, the planet doesn't need 'saving'. Our planet – Earth – will be absolutely fine. Whether you drive a car, or drill for oil, or watch literally hundreds of hours of television rather than getting on with writing your book, Earth will keep on spinning and spinning until the sun envelops her in about 5 billion years' time.* It's really just humans, plants and animals that need to be worried . . . though, if we're being completely honest with ourselves, animals are only in this mess in the first place because of us (except, possibly, for Peppa Pig, who has a real pro-fracking energy).

So, 'twelve years left' is a bit of a red herring,† but the science is pretty clear on what it means: 'for humans and animals to survive and thrive, we need to reduce global greenhouse gas emissions by 45 per cent by 2030, compared to 2010 levels, in order to keep the global average surface temperature below a 1.5°C increase by 2100, compared to pre-industrial levels'. Not quite as catchy, and almost impossible to fit on a T-shirt.

---

* A bit like what Amazon are doing to every other company in the world.

† Bad news for herring lovers, as warmer oceans mean they are impacted too.

In fact, this is a good example of why communicating climate change is hard, and one reason why this book exists. I don't know how many you've read, but most books and reports about climate change are, to put it politely, not particularly accessible and sleep-inducingly dull. It's all 'Look at how bad this chart you don't understand is,' or 'Look at this skinny polar bear, even though you have never met a polar bear, so don't actually know how skinny or not they're meant to be.' Well, like an old-timey explorer with a museum wing to fill, I say – stuff the polar bears! In my humble opinion, all reports on climate change should be combined into one massive poster that says 'We've got a hell of a lot of work to do in the next ten years, otherwise things will definitely start to get really bad, really quickly, for all of us and our lives will change forever.' Again, not ideal for a T-shirt, but I'm a climate-change economist, not a professional T-shirt slogan writer.

So – are we screwed?

The answer is not binary. People either want me to say 'no' so they feel better, or they want me to say 'yes' so they can give up. The answer is neither. The answer is to get involved to nudge the dial because we are not screwed yet, but the only way to make sure is to be part of the solution. I will say this: the world, as we currently know it, is coming to an end, regardless of whether you want it to or not. Either we do nothing, continue as usual, and the planet continues warming beyond recognition causing untold human misery, *or* we make some big decisions, work together to create a brand new, low-carbon society, and the planet is roughly recognisable.

That is the fight on our hands; the decision to be made. Do we want suffering or effort? Hardship or upheaval? Those are the difficult choices we face. Either way, change is coming. At least with the latter, it is within our control. The Chair of the Intergovernmental Panel on Climate Change (IPCC), Hoesung Lee, put it perfectly when he said in 2018: 'Every bit of warming matters, every year matters, every choice matters.'[1]

4

What is the climate and why is it changing?

For most people, knowing about climate change is a bit like knowing about your partner's job: you're pretty sure you understand what they do every day, but as soon as you have to explain it to someone else, you suddenly have absolutely no idea if they work in a bank, or actually just went to a bank once, and are in fact a photographer.[*]

And that's nothing to be ashamed of. For instance, my dad has a son that works on climate change,[†] but a couple of years ago when I asked him if he knew what I did, he replied: 'Yeah, climate change – it's getting hot and it's something to do with the ozone layer,' which starts well and ends incorrectly.

---

* This sounds unlikely, but happened to me, and it took almost three years of terrible financial advice and absolutely stunning Instagram pictures to resolve.

† Me.

Luckily, you don't need to be a genius – or really need to know anything – to enjoy and understand this book, because we're going to start with the basics. When we're talking about climate change, we don't begin with opinions, or feelings, or magazine horoscopes. We start with – to paraphrase that Jennifer Aniston hair commercial – *the sciencey bit.*

Weather vs climate?

Put simply, climate is all of the weather over a long period of time, normally thirty years. Sometimes you hear people saying things like 'If they can't predict rain on Tuesday, how can they predict climate change?', which sounds logical, but is actually complete nonsense. Because climate is about long-term averages.

Weather is the day-to-day changes that are reported at the end of the news, or – if you're more old school – that you see when you open your curtains. Despite the best intentions of weather-forecasters, the small, quick shifts of weather can be quite chaotic and unpredictable from day to day, hour to hour and even minute to minute, as you'll know if you've ever been caught in the rain.* But, unlike weather, *climate* is predictable, because small shifts average out over time. This is how we know that, in the UK, the first day of January will probably be colder than the first day of July.

So, when some Smart Alek says the fact that it's snowing means there can't be global warming, it's like seeing a sumo wrestler jumping on a trampoline and deciding that gravity

* Drinking piña coladas optional, but preferred.

doesn't exist. Just because he's going upwards *some* of the time, doesn't mean he can fly like a twenty-stone Peter Pan. That's why, when we talk about the climate changing, we are specifically talking about how climate over the last few decades compares to longer-term averages, like over the last 200 years.

You can also think about it in terms of a relationship. Weather is the day to day, up and down of long-term companionship. Some days you argue, other days you can't stop laughing. You're watching TV one minute, planning a heist the next: it's a real pick-n-mix. *That's* the weather. Climate is the long-term stable foundation upon which that relationship is based: your shared values, your history, the knowledge that neither of you can be bothered going to the gym to get fit.

But climate *change* is like finding out that your partner has, all of a sudden, completely out of the blue, started really getting into CrossFit. It's unbelievable. At first, it's just a class a week, but before you know it, they are getting up early every day to head off to 'the box' before you've even woken up, and they come back talking about how they could 'feel the burn' today. Soon they speak non-stop about their new CrossFit friends Murph and Fran. When you're out and about they are pointing out things they could 'deadlift'. All these awful protein shakes are kicking about the fridge. There's lycra everywhere. After a few months, you no longer recognise the person you agreed to spend the rest of your life with. It's in sickness and in health, not in fitness and in health. Then you find out their CrossFit instructor, Vincent, is giving them private lessons. Vincent is about ten years younger than you and in much better shape. You both bump into Vincent in Sainsbury's and it's awkward. But you put it to the back of your mind. But, when they finally leave you and move in with

Vincent, you start piecing it all together, and realise how long it has been going on, right before your very eyes.

That's climate change.

Except nobody is getting buff.

This story may or may not have happened to my mate Ian, who is now newly divorced.*

Like the weather – which can be sunny in Margate, and windy in Weston-Super-Mare, and sunny and windy and f*cking freezing in Bognor Regis, all at the same time – local areas also have their own climates. Which means that humans that live in that place have built their society, and their wardrobes, around a certain lifestyle. For instance, in Glasgow, we expect it to rain. Rain in Glasgow is like wine in France.† Because it's rainy and usually cold, buildings in Glasgow don't have air conditioning, and we don't own appropriate summer clothing, so if the weather does briefly get warm, all we can do is take our *taps aff* (tops off). My point is, our clothes, buildings, infrastructure, transport, etc. – basically, our lives – are completely geared around what Glasgow's climate is like: *dreich*.‡ So, take Glasgow's average rainfall in February over the course of thirty years – that's climate. And we can average up all February rainfall in places around the world to understand climate on a global scale.

The funny thing is that, as Brits, moaning about the weather

* In Ian's ex-wife's defence, he is a bit of an arse and probably drove her into Vincent's strong arms by getting her vouchers for CrossFit because she had 'put on a bit of timber over Christmas'.

† If you have too much at lunchtime, it's an unwritten rule that you get to go home and nap for the rest of the afternoon. Actually, I don't know if it's a rule, but I've always done it anyway.

‡ Scots, adj. tedious, wearisome; damp, wet, grey weather.

is what brings us together. It's our national pastime, overcoming all differences: the UN of conversation topics. But talking about climate is much less common, and only seems to drive us apart. And these long-term weather patterns (climate) are changing.

The greenhouse effect

The Earth's climate system, like me after a big night out, is very fragile and delicately balanced. But instead of being made up of six pints, a kebab and about forty minutes of proper sleep, it is made up of five components: our *atmosphere*, which is the gases surrounding the planet; the *hydrosphere*, which is our oceans and rivers and puddles; the *cryosphere*, which is all the ice and snow, and sounds like a planet that Superman comes from; the *lithosphere,* which is too hard to say if you have a lisp but refers to the earth's crust; and finally, the *biosphere*, which is everything that is alive, e.g. Joe Biden (currently). All these elements of the planet swap and transfer energy in such a way that there is a balance. And Mr Miyagi was right: the most important thing is balance.

At its simplest, the amount of energy in the climate system is set by radiating as much energy back into space as it receives from the sun. About a third of the sun's energy is immediately reflected back out again into space. The rest reaches the earth's surface and is converted from light to heat, warming the land, the sea, and you (if you're sunbathing). However, the very act of getting hot then emits its own energy (infrared radiation), which escapes back into space. Infrared radiation is invisible to us day to day, but it is measurable and observable. In fact, if we all wore infrared goggles all the time, then we would be looking at climate change in action, and literally nobody could say it's not happening . . . but I get it, they're not 'in fashion' right now.

How much energy the earth keeps or sends back is regulated by many things: variations in the amount of energy we receive from the sun, changes in Earth's reflectivity due to clouds, volcanic eruptions or snow cover, and last – but certainly not least – greenhouse gases. These are water vapour, carbon dioxide, methane, nitrous oxide and CFCs, in order of importance. That's right, the stuff you heard about in A-level chemistry and assumed would never have an actual impact on your life. Maybe we should all have been paying a little more attention to boring Mrs Hill, the dull Chemistry teacher. If only she hadn't had such a monotonous voice the planet might not be going to hell in a toasty handbasket.

These gases can absorb some of this infrared radiation. This is the Greenhouse Effect. Even though nobody under sixty has owned, or even seen, a greenhouse for over three decades now. These gases can play a huge rule in preventing the infrared energy from leaving. It could also be called something like 'the burning blanket' effect, because that's really what it's doing: wrapping a warm blanket around the planet that keeps in heat.

Beyond the name, all you really need to know is that the more greenhouse gases are emitted, the less radiation can escape back into space, and the warmer the planet gets. Simples. The natural greenhouse effect means the earth's surface has an average temperature of about 15°C. Without these gases it would be around minus 18°C, and we would live in a very different world.[1] In fact, different worlds do have different atmospheres. The Earth's atmosphere is 0.04 per cent carbon dioxide, whereas on Venus it's more than 96 per cent, hence why its average surface temperature is over 400°C.

History

If this all rings a bell, it's because we've known about the greenhouse effect in one form or another for absolutely ages. Back in the 1820s, the French mathematician Joseph Fourier was one of the first to consider how the earth's energy worked. This clever chap worked out that the earth should be way colder if it wasn't for *something*, although he wasn't sure what. As we now know, he was bang on, as without the naturally occurring greenhouse effect, we understand that the earth would be much cooler, and the fashion world would have one look all year long: winter coats. This means that, without *some* greenhouse gases, humans would be unable to evolve, survive and make banging techno music.

And that's not all! The 1850s were actually a ripe time for burgeoning climate research. An amateur scientist in the US called Eunice Foote undertook some experiments – she essentially left jars with moist and dry air, as well as carbon dioxide, out in the sunlight and measured their temperature. She found the jars with moist air (i.e. water vapour) and carbon dioxide warmed more than dry air. Though her lo-fi experiments were unable to isolate the greenhouse effect, the paper she wrote in 1856 made her the first person to hypothesise that the earth might be warmer if there was more carbon dioxide. Foote's paper was presented at the tenth annual meeting of the American Association for the Advancement of Science in New York. As a woman, however, she was not allowed to be present herself, so her paper was read out by the eminent American scientist, Joseph Henry. He was an advocate for women in science, and in his talk said, 'science was of no country and of no sex'.[2]

Three years later, in England, a lad called John Tyndall did some more concrete laboratory experiments by building some kit to measure the exact absorption of energy by the gases. He discovered exactly the same thing: the powerful absorption of greenhouse gases. He was then credited as having discovered the effect, and Eunice was forgotten until someone accidentally stumbled across her in 2011. To be fair to Tyndall, it seems he was unaware of Foote's work.

A decade later, in the 1860s, an interesting discovery about the role of the planet's orbit in causing natural climate change was made at what was to become the university where I did my PhD in Glasgow. But this discovery was made not by a professor, but by the janitor.* That's right, in a real-life period piece version of *Good Will Hunting*, the self-educated James Croll, who came from a Perthshire farming family, would stay after work to read books in the library. There, he came up with a theory about ice ages and how the earth's orbit caused climate variations. This was about fifty years before the more widely known Milankovitch theory on the subject. Croll then sent off a scientific paper, despite having only been educated to the age of thirteen and never having written an essay before. The cheeky rascal signed it off as 'James Croll, Anderson's Institution', and needless to say the scientific community was shocked beyond belief when they later found out he mostly cleaned toilets. But it led to him joining the Scottish Geological

* Or, as we say in Glasgow, the janny.

Survey and being awarded an honorary degree by St Andrews University – back when that still meant something, and they weren't given out to TikTok influencers for services to 'being an absolute legend', or however it works these days.

In the 1890s, the Swedish chemist Svante Arrhenius became the first person to look at how carbon dioxide may have affected the ice ages. He calculated that if you halved levels of CO_2, then the temperature would drop by 5°C. One of his annoying work colleagues came round and was like, 'Have you thought about this, blah, blah, blah,' and so, to shut them up, he did the calculation again for what would happen if CO_2 levels were doubled . . . and it was basically the same, but in the opposite direction. Considering that these calculations from the late 1800s were done before we had Microsoft Excel, they're not far off at all, and modern models suggest a temperature rise of somewhere between 1.5°C and 4.5°C for a doubling of CO_2.

Then, in the 1930s, Guy Callender – whose name sounds like something a saucy grandma might hang behind her kitchen door, but who was, in fact, a British engineer – suggested that greenhouse gases that came directly from humans burning fossil fuels might be to blame for the fact that the planet appeared to be warming. The works of both Arrhenius and Callender were mostly ignored for various reasons, including the assumptions that the oceans would soak up all the emissions, that the CO_2 tested in the labs at sea level would behave differently to CO_2 up high in the atmosphere, and that there were too many other natural influences – but they were also ignored due to a general lack of interest, the poor science of the time, and the fact that, for ages, computers were absolute garbage.

As technology improved, however, so have our experiments and our understanding. In fact, in 1956, Gilbert Plass predicted that we were turning up the toaster dial on old Mother Earth to almost exactly the temperature rise that we have now experienced, and the Scripps Institution of Oceanography hired Charles David Keeling to start taking measurements of the amount of carbon dioxide in the atmosphere. The first readings were taken in 1958, and they have shown an upward trend every year since. Therefore, the science is sound, and we've had plenty of warning, but, to put it bluntly, we've done sweet fuck all about it.

Where we are today

Today, the concentration of carbon dioxide in the atmosphere is the highest it's been in at least 800,000 years. Humans have only existed for a quarter of that time. Therefore, the very air you are breathing in now is the most carbon dioxide-rich any human has ever experienced – delicious!

'But how exactly did we release so much carbon dioxide?' I hear you shout at this book.

Well, what we've been doing ever since the Industrial Revolution is digging up lots of squashed plants and setting fire to them. Originally, we used dead-stuff chunks – which you might know as 'coal' – to help propel us in a big steam train from London Paddington to Bristol Temple Meads (before you had to go via Didcot Parkway), or we burned it in our homes to make one room in our house far too warm and clog our lungs. Nowadays, though, we mostly burn dead-stuff gas to make *all* the rooms in our home far too warm, or oil to help propel us along the M25 in an average amount of time. We use it to keep our computer servers cool so we can sit in silence with our partners and stream the entirety of *The Office* for the sixth time in order to dampen the irresolvable chasm of collective loneliness that modern society has created. Don't worry, the carbon footprint of Netflix and Chill is actually pretty small, certainly less than going for a night out.[3]

Over the last 800,000 years, the average CO_2 concentration has been between 170 to 290 parts per million. That refers to how much carbon dioxide there is per million parts in the atmosphere. The lowest readings were during the ice ages, when the temperature was 5°C colder. Throughout that time, they stayed in this range. However, when the aforementioned Charles Keeling first measured CO_2 concentration in 1958 at Mauna Loa in Hawaii,* it was already up to 315 ppm.† By 2015, the annual average was above 400 ppm for the first time in several million years.[4] Like a balloon slipping out of a child's grasp, CO_2 measurements are just going up and up and up, faster and faster – and only a shitload of human ingenuity is going to bring them back down. Longer records mean we can estimate that the last time CO_2 concentrations were this high was over 3 million years ago – back when you could only get four TV channels.[5]

If you had any lingering doubts about the correlation between changes in CO_2 concentration and global temperatures, they're all laid out clearly in the historical records: and, boy, oh boy, have temperatures been rising. 2020 was the joint hottest year on record – tied with 2016. (It is also my PIN. I always change

* Because if you are going to research the end of the world, you might as well do it in grass skirts and with a cocktail, right?

† Measurements are taken there because the air is so clean, away from any pollution.

my PIN to the warmest year on record. Am I more susceptible to fraud? Absolutely, but the message is getting out there, and that's the important thing.) The last six years have been the six warmest years on record. The last three decades have consecutively been the warmest decades on record. Across the world, we are seeing this. 2020 saw records broken for the warmest annual temperature for forty-five countries.[6] In the UK, the ten warmest years have all occurred since I was seventeen years old, while none of the ten coldest years have occurred since my parents were children.[7]

But that's not even the only marker: there are so many other signals and evidence that the planet is changing in line with our expectations about how greenhouse gases and temperature interact. We're seeing rising average sea levels across the world and warming oceans. We see that glaciers and ice caps are melting faster: in 2020, we saw the second lowest amount of Arctic sea ice during the summer.[8] Plants are growing earlier in spring, bird migration patterns are changing, and some plants and animals are shifting where they live.[9]

How do we know for sure it's us?

There is of course a certain amount of natural variability in the global climate as we've seen from ice ages. So how do we know for certain that we're causing this climate change?

When it comes to climate change, the specific type of change we are seeing now hasn't happened naturally in the past. In fact, all the data points to a change that natural reasons alone cannot explain. The global temperature has already increased by over 1°C since the Industrial Revolution, which is a period of just over 150 years. This is ten times faster than coming out of the last ice age, when, even at its fastest, it took about 1,000 years for the earth to warm by 1°C. And we're not even coming out of one. In fact, we're due an ice age, but human actions have probably already delayed it by 40,000 years.[10]

Also, if it was the sun causing global warming, then the outer atmosphere of the planet would be warming.* But it's not. The stratosphere is cooling. It is only the inner atmosphere that is warming, which points to the fact that it's the reflected energy coming back up from the earth, not energy from outside, that is causing all the bloody warming, and this is consistent with the greenhouse effect.

Asking 'What about natural causes for changes in climate?' is obviously fair enough. The annoying thing is that when someone says this, they usually say it as if they are the first person to ever consider it. It's crazy to assume that the scientific community hasn't thought about natural climate variations, and that specific disciplines haven't been studying them for decades. *As if* the Intergovernmental Panel on Climate Change doesn't explicitly cover this in a chapter of their reports.[11] No, no. It's definitely you – a balding man in the front row of my show, wearing an Anthrax T-shirt and sitting next to your wife, who clearly hates being seen in public with you – *you* are the originator of this idea. A modern-day Galileo, indeed.

The logic here is essentially that because something caused something in the past, it must be causing it now. You don't need to be Einstein to know that this logic does not hold up

---

* The inner layer is the troposphere, and the outer layer is the stratosphere – you'll have heard of that one because that is where my career is not heading.

to scrutiny. Sure, last time Colonel Mustard killed the victim in the Billiard Room with the candlestick, but you can't just assume it's the mustard man again. That isn't how you play *Cluedo*. Look at the evidence and be a damn detective!

In fact, all these other natural factors taken together are probably slightly *cooling* the planet. The best guess is that humans are actually causing about 110 per cent of all the warming.[12] And before some person smugly says, 'You can't actually give 110 per cent' – well, it turns out you can. And you need to stop doing it.

Every single way we look at it, the evidence now points to the fact that the planet is warming, climate is changing, and that we are causing it.

5

Twelve-week scan

You are stronger than you ever thought possible

I saw our baby for the first time on my birthday. Witnessing them on a screen was easily the best birthday present I had ever received, and I once got a DVD of *Transporter 2* signed by Jason Statham. And, if I'm honest, this was the first occasion I had got emotional about the pregnancy. My wife was mostly just relieved when she could empty her bladder – and to hear that it wasn't twins. I didn't think I'd be thinking about climate change at the scan. But I was, because I was sneezing a lot throughout from hay fever. It was late February. For the first time, I had hay fever on my birthday. I mean, as far as climate change impacts go, it's pretty low on the list. I feel stupid complaining about dry eyes and sneezing when others are suffering from floods and illness. But for me, it is a sign of what danger is to come, like the water trembling in *Jurassic Park*.

I have never really been a birthday person. My wife is, and I love that about her. It took me a while to get on board with birthdays as being worth celebrating, but now I am mostly a convert. Today, I didn't need to try. As well as filling up on water and having her tummy covered in cold jelly to be pressed on hard by a sonographer so that we could see our baby, my wife also got me a *Mandalorian* T-shirt, and a Baby Yoda baby-grow to go with it. If you don't understand what I just said, it is to do with *Star Wars* and it means we are cute. After the scan, we went for a meal with some friends and told them the good news. We then spent about an hour walking about London, trying to find a late-night karaoke bar. My mate Ian gave an incredibly heartfelt rendition of Take That's *Back For Good* accompanied by real tears – every single time it was his turn. On the way back, sitting on the Tube, we both stared at the picture of our tiny new person.

I'd spent the first few weeks after we found out we were having a baby in a state of understated happiness and quiet shock – and then my instincts kicked in. Everything up to this point had felt hypothetical, but now I needed to find out more about how to prepare for a birth. This involved finding out what the hell hypnobirthing was, as my wife mentioned that it had been recommended to her by friends. I was fairly sceptical, as it sounded like Derren Brown might be at the birth, but I thought it best to keep an open mind and be supportive. And shit, we needed to move out of the one-bedroom flat we'd happily been renting for years.

And where would they be born? Thankfully, once again, my wife is also excellent at preparing for and researching stuff.

At the same time, my academic research brain swung into gear. I resolved to find out more about birth, babies and climate change, and tried to make sense of what was happening and what would happen. Maybe if I could get enough of a grasp on all this, I could be more certain of what lay in store.

I knew my child would be born into a world that has already altered from the one I was born into. It is likely my child will only see coral reefs on history documentaries. What might the future look like for a baby born today? It is extremely uncertain. It depends on what we do. It also depends on where you are born. It is likely that, without a radical decline in emissions over the next decade, we will see the world pass a 1.5°C temperature rise by the time my child turns twelve. Some irreversible change is certain. I don't want to look back in twenty years' time, when I have hay fever at Christmas, and wish I had done something sooner.

As a climate researcher, I also knew that there are always two sides to every climate coin. There are the impacts of climate change on a child's life, and there's also the impact a child will have on the climate. In other words, does having a child *cause* climate change?

I remembered something I'd seen someone post on Facebook a few years earlier: 'Want to fight climate change? Have fewer children'. It was a *Guardian* headline from July 2017, accompanied by a picture of three little babies all looking equally adorable and gormless. The tagline said, 'Can you bring yourself to have one fewer of these?' The article claimed that the most impactful thing an individual can do to fight climate change is to decide to have one child fewer. I had only just decided to have *this* child. If I had one fewer, that would be no children. Was having children really that bad?

I put on my researcher hat and read the study the article was based on.[1] It considered what the most effective actions were in terms of reducing an individual person's CO_2 impact.

And found that the things you're often told to do for the planet don't actually make much of a difference. Instead, the authors concluded that the top four things you as an individual could do were:*

In at number four: Bart Simpson was right all along. Don't have a cow, man. Or, at least, have less of a cow, man. It's 'Eating a plant-based diet'. Not eating meat and dairy saves about 0.8 tonnes of CO_2e (carbon dioxide equivalent) a year, which is over 10 per cent of the average UK person's greenhouse gas emissions in 2018.[2]

At number three: is it a bird? Is it a plane? Yes, it's a plane – and flying in one is the most emissions-intensive activity you can do. Taking a round-trip transatlantic flight emits 1.6 tonnes of CO_2e in only sixteen hours. You would have to go vegan for two years to save the emissions caused by going on a weekend break to New York from London.

Screeching in at number two is driving. Beep beep! For the average drive, going car-free would save 2.4 tonnes;[3] frequent drivers can save more than four times that much.[4]

And at number one? 'Having one fewer child.'

They also identified two other potential high-impact actions – not owning a dog and purchasing renewable energy – but didn't include these due to uncertainties about data.

Now, at this point, I had tried going vegan by doing Veganuary. I found it tough, though not as tough as calling it Veganuary without wanting to punch myself in the face. I'd cut down on my meat intake a bit since then. When it came to flying, as a last hurrah before trying to start a family, we had gone on

* NOTE: This bit works best if you hum 'Whole Lotta Love' by Led Zeppelin (the *Top of the Pops* theme song) in your head as you read it.

a big trip to Japan. There, we got to experience cherry blossom season, which I ruined by talking about how it was arriving earlier and earlier each year (in between sneezing). Afterwards, we'd basically stopped flying, and certainly wouldn't be joining the 1.60934 km high club, at least for a while. And at present, I didn't own a car.

But the last one – well, it didn't sit right with me. Is the best thing you can do for your child's future really to not have a child? People's emissions vary massively by wealth, geography and consumption – so one single number of an 'average' child seemed strange. Plus, it was too late to do anything about that now. I resolved to look at this in more detail and soon wondered, jokingly, if it was possible to offset having a baby.

6

Heatwaves, droughts and hunger. Is this some biblical shit?

'It's getting hot in here, so relocate your homes.' So goes the famous song by the American rapper Nelly. He is right: it's getting *scorching*. And, in the future, what you think of as the hottest heatwave you've experienced – that will just be a normal summer.

Heatwaves

Summers are getting longer. The hot days are becoming hotter and staying hotter for longer. On paper, a longer summer sounds better; who doesn't like summer? But longer isn't always better. Think about films. Ninety minutes is the right length for a film, and every minute over is a minute I wish I was peeing out the £8 Coke Zero I downed during the opening titles.

The UK, in particular, is not designed at all for warm weather. And we are doing next to nothing about it. Humans tend to build stuff based on our past experience. That's a problem, because the future will not be like the past, and things are hotting up, and our houses, workplaces, care homes and infrastructure are not prepared for this increasing heat.

While heatwaves can seem fun for tabloid newspapers who splash pictures of frolicking young women in bikinis eating ice creams across their front pages, the reality is that, for many, heatwaves are an absolute melting bastard. The heatwaves in 2018 saw 1,700 excess deaths in the UK – a country where the record temperature is 38.7°C.[1] So, still under 40°C [2] – for now.*

In fact, the chance of a summer heatwave, like the one we saw in 2018, has doubled in recent years in the UK.[3] Without more action, we will see 7,000 deaths a year in the UK from heat stress.[4] An ageing population means that overheating in care homes and elderly accommodation requires urgent attention. Just like with COVID-19, it's vulnerable people – the elderly and those with pre-existing conditions – who are more likely to suffer negative effects, such as dehydration and heat stroke. Globally, there has been a 54 per cent increase in heat-related deaths in people older than sixty-five since 2000.[5]

In 2021, a deadly North American heatwave saw many deaths as a heat dome enveloped a usually cool region, and officials observed Canada's previous high temperature beaten by nearly 5°C. It occurred in the tiny mountain village of Lytton, which registered 49.6°C.[6] This is essentially unheard of that far north. Wildfires then consumed the village.[7] In Salem, Oregon, it reached 47°C, a full 9°C hotter than the previous temperature

* And summer 2020 saw a total excess mortality comparable to that observed in England during the 2003 pan-European heatwave, in which 2,234 excess deaths were observed (that is after accounting for COVID deaths).

high. This heatwave event was made 150 times more likely to occur due to climate change.[8]

Heat island

The longer they last, the more problems heatwaves cause. and climate change is increasing the likelihood of strings of very hot days happening.[9] Heatwaves also cause issues when minimum night-time temperatures rise. Night-time should be when everything and everyone gets some respite from the punishing heat. This is a particular issue in cities, which experience even warmer temperatures due to something called the 'urban heat island effect'. This is where dense urban populations retain more heat energy than rural areas, because, it turns out, buildings and pavements have different properties to plants and manure. Things could have been different if they'd built London or Manchester out of manure and dandelions, but they didn't, so we're stuck with how they are.

With Heat Island, much like *Love Island*, things are steamy, you're probably going to feel uncomfortable if you are with your parents, and too much makes you feel like your brain has been hollowed out with a spoon. In summer, New York City is 4°C warmer than surrounding areas.[10] And more and more people have been moving to cities. (Except me. It's still stupidly warm here in the leafy suburbs, but thankfully we have escaped the balmy London nights.) During the 2020 August heatwave, London had many tropical nights where temperatures didn't drop below 20°C.[11] Goodbye! I do not miss those nights of sleeping naked with the windows open while hearing foxes furiously going at it outside my window. Thing is, though, I now can't get to sleep *without* that noise, so I've had to download a special app with a looped eight-hour recording of the sound of foxes sexually assaulting each other. Bliss.

The heat island effect is often strongest on calm nights, when there is no breeze to sweep away the trapped heat. Greening cities is one way to overcome this effect. Knocking them down and rebuilding them out of shit and dandelions, as mentioned before, would also work, but no politician has got the guts to campaign with that as their policy. Cowards.

The ability to cool down at night and regulate body temperature while asleep is crucial, and more medical issues arise when this is not possible. According to the World Health Organization (WHO), extreme heat can exacerbate pre-existing health conditions like respiratory diseases, heart conditions and kidney disorders. All the good ones, basically.[12] Less serious, but still important, is that a bad night's sleep caused by high temperatures can also lead to an increased prevalence of passive aggression between couples. That's not from the WHO, but my own research.

Air pollution plays a significant role in the way a heatwave can affect health. This provides yet another route for cities to become dangerous places. Around 20–40 per cent of the deaths during the 2003 European heatwave can be attributed to worse air pollution.[13]

Heat records

A number of other records were also smashed across the world in 2020 – and not the fun type that Guinness are interested in. But, OK, to keep things light, I'll throw in one fun one, so keep an eye out for it:

- The highest temperature ever recorded by humans on earth – 54.4°C – was registered in Death Valley during August 2020.[14] (This was beaten, again in Death Valley, in July 2021 by 0.1°F.[15])
- The hottest ever January, May and September worldwide.
- On 4 January 2020, it was 48.9°C in western Sydney, the highest temperature ever recorded in an Australian city.[16]
- Baghdad hit its hottest ever temperature – 52°C.
- In November 2020, Martin Rees from Hertfordshire performed twenty magic tricks underwater in three minutes.
- 2020 was the third-warmest year in the UK, and the UK's top ten warmest years have all occurred since 2002. Temperatures of 34°C have been recorded in the UK in seven out of the last ten years (before this, we only saw 34°C reached in seven out of the previous fifty years).[17]
- Temperatures in Siberia were 5°C above average during the first half of 2020, and reached 38°C in the Russian town of Verkhoyansk in June, the highest temperature ever seen north of the Arctic Circle.[18]

Not sure if you noticed but that last one was the *Arctic*, by the way, and if it's been hard to contextualise temperatures in your head, then surely this one should stick out as a red flag. This Siberian heatwave was made 600 times more likely by climate change.[19] Watching *The Snowman* at Christmas will be a lot more harrowing if they venture north to find Santa and all the other Snowmen engulfed in an inferno.

And it's not just 2020: July 2021 was the warmest month ever recorded worldwide.[20] 2019 broke more records than a sweaty-palmed vinyl enthusiast, including the hottest temperature ever recorded in the UK on 25 July 2019 in Cambridge – an event that was made twenty times more likely due to climate change.[21] We also had the UK's hottest ever winter temperature that year, during the warmest February on record.[22] (I haven't checked how many magic tricks were performed underwater in three minutes that year.)

The only upside I can see is that siestas will probably become a traditional part of life for everyone in the near future.

Around the world, extreme heat affects our infrastructure, as it melts roads, buckles railways, causes power outages, and grounds planes.[23] And all of this is happening now, with only around 1°C of warming so far. Things will undoubtedly get hotter. What the future looks like obviously depends on what we do, how much we do and how quickly we do it.

Too hot to handle?

Will parts of the world soon become too hot for humans to live in? The answer requires a quick explanation of something called 'wet-bulb temperature'. The temperature we see on the weather reports on the news, and that we all use, is called the 'dry-bulb temperature'. This means it doesn't account for humidity. Once the dry-bulb temperature gets above about 35°C, we start sweating to cool down. Stinky armpits, yes, but we know what we're dealing with, and humans can physically cope.

The wet-bulb temperature includes humidity, and so is a lot higher in real terms. Basically, it covers the important detail that sweating won't cool us if there's too much moisture in the air. To measure it, you cover a thermometer in a damp sock or rag and swing it around. I remember a science teacher from my school near Glasgow telling us about this, and a number of my more badly behaved classmates suddenly started taking an interest, mostly because it sounded like a novel way to attack someone.

And when the wet-bulb temperature reaches 35°C it becomes so hot that sweating no longer works, and the body can no longer cool itself. Without relief from this heat, we will start cooking from the inside out, like an over-microwaved sausage roll. (On the plus side, I guess, if you've ever wondered whether climate change might lead to you being melted to death, you don't have to worry – the humidity will kill you first. It's a bit like how I'm comforted by the fact that, in a housefire, you're more likely to die from smoke inhalation than being burned to death.) We have begun to see flickers of the wet-bulb temperature reaching such levels in the hottest places on Earth, such as the United Arab Emirates.[24]

Central estimates of where we are heading at present suggest an overall temperature increase of around 3°C by 2100. While this isn't enough to bake us alive, it will still be an absolute shitshow. People can't continue with outdoor activity if the wet-bulb temperature gets above a certain level – which would be the equivalent of about 55°C to our dry-bulb understanding of temperature. Working outdoors during the summer will become impossible in many parts of the world. 'Sorry I can't come to work today, the planet is on fire' sounds like a decent enough excuse – although, obviously, it's going to happen in places where *not* going in to work is not an option. Half of the world's food is produced by small farms that rely on physical labour.[25] In 2019, over 100 billion hours of work were lost to rising temperatures compared to the year 2000.[26]

It means people will be on the move as places become uninhabitable. Climate change is a threat multiplier. That means it can make dangers – things like conflict – even more dangerous. In fact, climate change is now considered a global security issue by none other than those hippy tree-huggers at the Pentagon.[27]

Droughts

Climate change is also making some droughts worse.[28] A drought can be defined in various ways: a lack of rain (a meteorological drought); effects on water supplies (a hydrological drought); or effects on soil moisture (agricultural drought).* Increased temperatures in summer are often accompanied by less rain, leading to drier conditions and droughts. Warmer temperatures also increase the amount of water evaporating out of the soil. Summer rainfall is likely to come in shorter, more intense bursts, which flow off very dry soil instead of being absorbed and might even lead to flash flooding.

Obviously, a lack of rain is different in Glasgow compared to, say, Bahrain. Droughts are therefore measured as a reduction against the average from that specific place. Sea-level rise can also cause drought, as the saltwater gets into the freshwater supply in underground aquifers. This reduces the amount of fresh water available – something that is already happening

---

* I've always admired meteorologists because they mainly just tell us if it's going to rain, but have chosen a title that's a bit like 'Asteroid person'.

in Florida – and makes soil saltier. It can even render coastal farmland infertile[29], as is happening in Bangladesh, where many farmers are being forced to turn towards fishing.[30]

Scientists have recently started highlighting that climate change is making specific droughts worse or more likely. Between 2011 and 2017, California had one of the worst droughts in the state's history, which killed 102 million trees.[31] Climate change has already increased drought risk in southern Europe, where water availability is significantly impacted by soaring temperatures.[32] In the 3°C future we are currently heading towards, southern Europe would be in a constant drought.

England is also expected to face more droughts within the next twenty years. The head of the UK Environment Agency, Sir James Bevan, has called for water use to be reduced by a third, and said he wants to 'make it as socially unacceptable to waste water as it is to blow smoke in the face of a baby'.[33] Which is, well, oddly specific. My main concern is that, in order to try to get around this, English people will simply try to make it *more* socially acceptable to blow smoke in the face of a baby.

Australia had its hottest day ever on 17 December 2019. The next day, it beat that very record by 1°C.[34] This gave the very small subsection of Australian society that are both meteorologists and *Crocodile Dundee* fans a chance to say: 'Call that an unprecedentedly high dry-bulb temperature? *This* is an unprecedentedly high dry-bulb temperature'. Meanwhile, the changing Australian climate means water supplies in parts of the country are becoming more dependent upon desalination plants, where the salt is taken out of sea water so it can then be used for drinking. Carlsberg Brewery are creating a desalination plant in India to create clean water for a West Bengal town of 4,000 people. Which is probably the best desalination plant in the world.[35]

A drought over a period of years meant that Cape Town in South Africa thought it would run so low on water in 2018 that it would be necessary to shut off the public's water supply, an event referred to as 'Day Zero': turn on the tap, and nothing comes out. Water use was restricted to 50 litres per person per day (for comparison, the average UK resident consumes 142 litres daily).[36] Day Zero was barely averted, and farmers in the area lost around a quarter of their crops.[37] This drought was made at least five times more likely due to climate change. Similar ones may become much more common in the future.[38]

Lack of access to water is already a massive health risk in many of the world's poorest countries. Approximately a quarter of the world's population live in countries that are water-stressed, and 785 million people do not have access to basic drinking water.[39]

Food security

Our food supply is vulnerable to droughts and extreme heat. The direct response of agriculture to higher concentrations of carbon dioxide in the air can be more complicated, as carbon also has a fertilisation effect that makes plants grow bigger (nature's Viagra). However, research shows that nutrients do not increase from CO_2 fertilisation, meaning the average nutritional content drops – this is a pretty important caveat, especially in developing countries. Basically, CO_2 fertilisation is the equivalent of steroids – a mug's game.

Regardless, it is clear that more predictable weather is preferable for farming. What we eat in the UK is already being affected. Due to the 2018 summer heatwave – which was thirty times more likely to occur due to climate change – our crops did not get the water they required to grow.[40] Due to this lack of water, chips in the UK the following year were an inch shorter.

That is not a world I want to live in. You can't dip wee chips. For God's sake, think about the condiment industry. One of life's only joys is finding a big chip in a bag of chips, and holding it aloft like you're in *The Sword in the Stone*, proclaiming to everyone else within earshot: 'LOOK AT THE SIZE OF THIS EFFING CHIP!' to which someone will always reply: 'IT MUST HAVE EATEN ALL THE OTHER CHIPS!' and you'll all roll about laughing. Well, your children won't have that. That 'big chip' experience. That is what we are fighting for here, people. Proper chips.*

Climate change is also going to reduce barley yields and raise the price of beer.[41] If the public can't rally behind big chips and cheap pints, then I don't know what'll solve climate change. Economies that are reliant on a couple of agricultural products will be most at risk. Coffee is another product being affected by climate change, and it is likely that scarcity will see the price of a cup of joe rise. A 2°C temperature increase would reduce the growing area of Uganda's Robusta coffee to only 10 per cent of its current size.[42] Which means it will probably have to focus on making espressos.

---

* And difficult choices will absolutely need to be made. You can either have big chips or you can keep flying to Magaluf to finger someone from Middlesbrough, even though you could just go to Middlesbrough. The solutions are in front of us, people!

Animals

Animals have it bad. They are often less able to adapt, and getting a pet passport is a hassle. The current climate conditions, which have been fairly stable for thousands of years, have allowed life as we know it to flourish on the earth. And we are changing that in the blink of an eye. Humans have some capacity for adapting, but most of nature cannot do so at such speed. Evolution takes time: it is almost as slow as getting through to Virgin Media's customer services. We are conducting the biggest survival-of-the-fittest experiment ever undertaken.

Except it's not really the fittest, is it, when you've got one paw tied behind your back? Many animals are shifting their territories towards the North and South Poles – reptiles, amphibians and insects at a rate of about 7, 12 and 18 metres per year, respectively.[43] I've already seen spiders in the UK that are so big, that for the first time in my life I have felt admiration for Australians.

Others are moving upwards to find suitable conditions. Mountain butterflies are climbing, and are no longer found at some lower altitudes.[44] At the same time, climate change is threatening animals in the cloud forest of the neotropics, where some unique species can only survive at certain heights. They are being pushed further and further up – until they risk running out of mountain. Although, really, by being mist-enshrouded magic lands with twisted, moss-covered trunks, the cloud forests of the neotropics *have* been tempting fate. If they want to be something out of a fairy tale, well, their wish might have come true.

But many animals can't move fast enough – or at all. The

first mammal to go extinct due to climate change was the Bramble Cay melomys, a rodent that lived only on an Australian island near Papua New Guinea.[45] It disappeared in 2016 due to sea-level inundation and storms. A recent study found that on a current trajectory by 2070, 16 per cent of the plant and animal species they studied would be extinct.[46] That was best case, it could be over half.

Heatwaves are getting hotter, droughts are getting droughtier, and our chips are getting smaller. Parts of the world may well become uninhabitable, crops will fail, and lots of workers will be unable to work under such conditions, which will affect health, livelihoods and economies. Not exactly a cheery end to this chapter, but there is no escaping that due to human-caused climate change, the heat is on.

7

Are we all going to live underwater?

One of my earliest childhood memories is watching my father drifting away to drown at sea.

We were in the Algarve on holiday, and were at a beautiful beach when my dad went out on a lilo and got dragged away by the current. There wasn't a lifeguard in the vicinity. Or if there was, they were bad at their job.* I stood on the sand with my cousins as my aunt and uncle paddled out to rescue him. It's not a particularly dramatic story, as eventually all three came back, completely exhausted, and the lilo washed up much further along the beach several hours later. Plus, my mother missed the entire thing, because my younger brother had to go to the toilet. But ever since then, the ocean and I have always

---

* It was the early nineties, and thus the heady days of *Baywatch*, which emphasised the importance of style over substance in the lifeguarding community.

had a mutual respect and kept our distance.* The lesson is, never mess with the oceans. Except, we are messing with them.

With all the ominous predictions about the impact of climate change, it's easy to assume that the future is going to be exactly like the film *Waterworld*, i.e., terrible and expensive.† But what actually lies in store for H_2O – which makes up over 70 per cent of the earth's surface? And did the Little Mermaid jump the gun a bit by getting rid of her tail in exchange for some legs?

The oceans are important – and not just as a big bath for sharks, but because they take in over 90 per cent of the extra energy caused by the greenhouse effect (source: *Heat* magazine). The oceans absorb much more energy than the air. Therefore, the big blue sea is the best barometer we have to measure global warming. The five hottest years for ocean temperatures have occurred since 2015, with 2020 being a record high.[1]

To prove this water vs air difference, here's a quick experiment you can try. Light two candles. Above one, suspend a

---

* Today, I find sunbathing on the beach glorious, but I've remained wary of the power of the waves: indeed, 'I do like to be beside the seaside' is a mantra I have taken to heart, with the emphasis on *beside*. Now I think about it, that song doesn't make much sense at all. They are already talking about the side of the sea . . . and then they want to be beside *that*? Two steps removed? So, standing on the side of that road that runs parallel to the sea, next to the run-down arcade and a shop that sells saucy postcards and assorted tat – *that's* where they'd like to be? Each to their own, but I wouldn't devote a whole song to it.

† Kevin Costner played a fish-man who could live in water. He was like the opposite of a fish out of water. He was a man in the water. Also, why do people say that someone was like a fish out of water as if that's a bad thing, when the only reason we are all here – evolution – is precisely *because* a fish got out of the water? It should mean that someone is a maverick that's changing the world.

balloon filled with air, and above the other, place a balloon full of water.* It'll take a lot longer for the water-filled balloon to burst, because the water is able to soak up more of the heat – but still not nearly as long as it'll take to explain to someone what the hell you're doing if they happen to walk in on you. (Spoiler alert: you'll try, but you'll fudge it and it'll become a running joke. For years afterwards, your nickname will be 'Balloon Guy', until you completely forget why. You'll move to a new town, start a new job, and life will be fine, until one day, an old friend comes to visit and tells your new work colleagues about 'Balloon Guy'. They'll chuckle politely and move on, or so you think – until your probation is over and they let you go, saying you're not a good fit, with the note 'weird balloon guy' written on your permanent records. On second thoughts, maybe don't do the experiment. It's not worth it.)

Global warming is causing the sea level to rise through two main effects: the melting of ice caps and glaciers (48 per cent), and the expansion of the sea as it gets warmer (39 per cent).[2] That's right: water enlarges when it gets hot. So, the next time someone says you've put on weight, just tell them the human body is mostly water, so you haven't got fat – it's climate change.

Ice, ice, baby

Some celebrities are real campaigners on the climate issue. Leonardo DiCaprisun is really into climate change. He made a film called *Before the Flood*, has his own foundation devoted to climate action, and was the UN Messenger of Peace with a

---

* Sorry, Mum, for stealing all your Jo Malone candles in the name of science.

special focus on climate. I mean, the only reason he got famous in the first place was due to an iceberg, so it makes sense that he's trying to save them. Plus, it's personal to him. We know for a fact DiCaprio cares about future generations: just look at who he dates.* But, ultimately, he is right to care about ice, because it's very important.

Snow and ice play an important role in regulating the climate, because they are white. (Typical! They're probably all middle-class men that went to Oxbridge, too.) If you've ever travelled to somewhere where the climate is very hot, you might have noticed that the buildings tend to be painted white. This isn't *just* for Instagram – it's because light colours reflect heat, keeping the buildings cool. Conversely, darker surfaces absorb more heat. This is called the albedo effect. As snow and ice around the world disappear, they leave behind darker surfaces, which then absorb more heat, like some sort of deadly game of peekaboo. In fact, a bit of ocean that remains unfrozen can absorb up to nine times as much solar energy than if it was covered in ice, which in turn makes the world hotter, melting more ice: it's a snowball effect. (I guess snowball is a poor choice of words, but you get what I mean.)

A recent study used satellites to look at 215,000 mountain glaciers, the Greenland and Antarctica ice sheets, various ice shelves floating around Antarctica, and sea ice drifting in the Arctic and Southern Oceans, which collectively made up an incredible collection of images that was nonetheless described by some experts as 'a little samey'. This study found that

---

* This is the opposite of my recently divorced mate Ian who appears to have accidently set his Tinder preferences to over 60 and can't understand why he's always matching with his mum's friends.

between 1994 and 2017, the world lost 28 trillion tonnes of ice,[3] the equivalent of a sheet of ice 100 metres high covering the entirety of the UK – an event that would still see the men and women of Newcastle refusing to wear more than a vest or a short dress on a Friday night. Much of this water is now filling up the oceans.

The Arctic, specifically, has been warming about twice as fast as elsewhere in the world. The Arctic sea ice tends to shrink to its smallest size in September each year. As of September 2017, it was only 65 per cent of the size it was back in the late 1970s.[4] (A similar thing has happened to Mars bars, but this isn't really the forum for that discussion.) Even if global temperatures only rise by 2°C, that will increase the probability of an ice-free Arctic summer in any given year to 16 per cent. That probably rises to 63 per cent with a 3°C rise, i.e. it will occur more regularly than not.[5] Worryingly, the rate of melting ice in the Arctic is getting even faster. I threw a 'worryingly' in there just to clarify that we definitely want ice in this scenario. I know your interaction with ice is probably to find it annoying, because it's always either settling on your car and making you late for work, or filling up your freezer and making it hard to get at your peas. Please forgive it these transgressions and try to see ice as the protagonist in this story. Between 1990 and 2001, the Greenland ice sheet lost about 34 billion tonnes per year on average. However, the rate of loss increased to around 247 billion tonnes per year between 2012 and 2016. Antarctic ice loss also increased by four times over the same period.[6] I'd say you can't ignore cold, hard stats like this, but the problem is it's not cold and hard: it's wet and watery.

This warming is affecting life in the Arctic. For indigenous communities, such as the city of Shishmaref in Alaska, the altered predictability of their way of life is having profound effects. Those who hunt on the ice are finding generations of knowledge are going out the window as seasons abruptly change. In many ways, it's a bit like my ability to programme a VCR using a VideoPlus number: it used to be an invaluable skill in our household but is now obselete. But unlike my VCR skills, this is a real threat to their livelihoods. And for coastal towns, the protection from storms afforded to them by ice is often no longer there. In certain Arctic areas, melting permafrost will damage building structures and cause natural landscapes to collapse.

Not to get all Attenborough on you, but while I don't really care about polar bears, it is worth mentioning them here. Obviously, they need ice to hunt on, and these longer periods without ice over the summer put more pressure on them to quickly get the food they need when they get the chance. It is estimated that by the middle of this century, polar bears' numbers may be reduced by 30 per cent.[7] Basically, they're being forced to get beach body ready against their will. Interestingly, grizzly bears also appear to be moving further towards the coasts, which is leading to an increase in interbreeding, giving rise to the cute couple name of 'Pizzlies'. I wouldn't want to be in the room when those two are mating. Also, why they'd be in a room, I don't know. I'm just saying if they came into this room right now and started making out, I'd mumble something about needing to go to the shops and leave.

Reindeer are also feeling the pinch, their annual herd migration across the land is occurring earlier and becoming more dangerous, and leading to starvation.[8] We should keep an eye on that, because if we end up with fewer than nine reindeer, it's really going to ruin Christmas. 'Now Dasher, now Dancer . . . er, that's it.'

Glacial retreat is another concern, even though that phrase sounds like a luxury holiday advert for one of those ice hotels. Glaciers provide the majority of fresh water for humans and animals on the planet.* These massive, slowly moving blocks of ice feed rivers, and entire settlements are based around their existence. A healthy glacier's ice mass either stays the same year on year or increases. Most of the world's tropical and mid-latitude glaciers have underlying health conditions and are – like Prince William's hairline – shrinking and thinning. Glaciers in the Himalayas are not only experiencing warming, they are also having black soot deposited on them from modern industrial development in Asia – and black, as we know, absorbs heat. The city of Lima in Peru, with a population of ten million, uses glaciers in the Andes for both hydro power and water supply, though it recently had to start switching to natural gas for energy security to make sure power stayed on. It is now causing climate change because of climate change.

I am also worried as a keen skier. Glaciers in the European Alps are expected to lose 50 per cent of their volume by 2050 regardless of how much we cut emissions.[9] Things will become particularly bad for lower altitudes, as snow cover below 1,500 to 2,000m is expected to drastically decrease.[10] Verbier in Switzerland has removed two of its lower chairlifts due to glacier retreat. I would be piste-off about the lack of a ski season: I am hoping that there will still be places left I can take my child to learn to ski that are accessible directly by

* Plus, we might run out of mints.

train from London. Seeing lots of tiny three- and four-year-old French children in their ski schools all waddling about is one of my favourite sights in the world.

Rising sea levels

So far, human-caused climate change has caused sea levels to rise by over 25cm on average since 1800, and the rate of the increase has been speeding up over the last few decades.[11] During the last 5,000 years, sea levels rose about 1mm per year.[12] We are now seeing rates that are at least as fast as 3.2mm per year, and this is accelerating – it was 6.1mm from 2018 to 2019.[13, 14]

Now, you may not think you already know much about how climate change and the oceans will impact the world in the future, but I'm willing to bet that many of you remember this statement made by three pioneering young male scientists nearly two decades ago:

> *I've been to the Year 3000.*
> *Not much has changed,*
> *but they lived underwater.*
> *And your great, great, great granddaughter,*
> *is pretty fine.*
> – Busted *

Now, let's take this statement one piece at a time. Actually, first off, these boys should've been given a Nobel prize for

* Or the Jonas Brothers, if you are a young person. And if you are older than fifty, then this next bit will go way over your head.

talking about climate change back before Al Gore, so good on them. I wasn't thinking about climate change back in the early 2000s. I was thinking about getting good grades in my exams. Not because I wanted a good job or anything, but because if I got good grades, my parents would let me and my mate Ian go to Magaluf for a week.*

Anyway, now I am older and have a PhD, I can peer review the comments made by the scientific institution known as Busted. What they do not specify are their assumptions regarding a global CO_2 concentration pathway and the associated temperature over the coming millennium. Well, maybe they do sing it during the middle-eight breakdown of the song, but I've never listened that far.

So, what does the science tell us about their claims? Well, they might not be far off the mark. Even in a best-case scenario, by the end of the century, the sea will be 30cm above the level it was in the year 2000.[15] As recently as 2013, the IPCC thought that around half a metre was likely on current trajectories, and that even in a worst case, it was unlikely sea levels would rise by over 1 metre this century.[16] However, more recent thinking suggests this could be a significant underestimation due to the behaviour of the Greenland and West Antarctic ice sheets.[17] NOAA (not the guy from the ark, but the aptly named National Oceanic and Atmospheric Association) recently revised its best- and worst-case estimates upwards. Due to deep ocean heat uptake and the continuing loss of the ice sheets, the seas will keep rising for centuries as we move

* I did get good grades, and we went – and Ian met someone very nice from Middlesbrough.

towards the year 3000 and will remain elevated for millennia, so who knows, Busted may well be correct?[18]

My second issue is with their assumptions about human adaptation to climate change. That we will simply live underwater. Sure, there are some small upsides – for instance, I will no longer need to listen to my stoner mate Mad Scoobie's stories about his trip to Amsterdam, because the Netherlands will be underwater. But mostly, it is not going to be great, and, on the whole, places won't cope. Four degrees of warming would inundate land that is home to between 470 and 760 million people.[19] Even with a rise of 3°C, flooding will reach 275 million people, including coastal cities such as Osaka, Miami, Shanghai, Alexandria and Rio de Janeiro. I mean, if we can't save the Copacabana, the sexiest place on earth, then what is even the point?[20]

Now, I know many people would like to see their government sent to the depths of the ocean, but in 2009, the Maldives government actually held an underwater cabinet meeting, spending thirty minutes on the seabed in full scuba gear, to draw attention to the country's plight due to global warming. Many such low-lying island nations are at the forefront of climate change. Places in the Caribbean Sea and the Indian and Pacific Oceans are already being terribly affected by sea-level rise. Island nations like Kiribati and Tuvula will no longer be *Pointless* answers – because they simply won't exist.

The impacts on these islands are severe. Coastal flooding is inundating their freshwater supplies and ruining crops. Food security is also threatened as reefs disappear. When the ocean displaces families from their homes, it can be tough to find new ones on the limited land. Displacement from encroaching sea levels can exacerbate ethnic tensions to do with land rights in places like the Solomon Islands.

To cope with these impacts, the islands can build new walls, plant mangroves, and raise up the stilts on houses even higher. The Marshall Islands are considering whether to create an entirely new island or attempt to raise an existing one, with both options being astoundingly expensive.[21] But adaptation can only go so far. Losses have already occurred, and damage has been done. These places are the least responsible for the problem that is now threatening their homes, their histories and their cultures. They need help.

The UK is affected too. The town of Fairbourne in north Wales is to be decommissioned due to predicted sea-level rise, which will make its inhabitants some of the first to become internal climate migrants in the UK.[22] Already the value of their homes has dropped so much that it is impossible to sell them in order to buy new homes elsewhere. The community is devastated. Large parts of the coast and farmland of Lincolnshire in England will also be completely flooded if temperatures rise by 3°C.[23]

My final issue with Busted's analysis is their demographic assumptions around life expectancy. I've done the maths, and, assuming they were talking to someone my age when the song came out, then in the year 3000 my great, great, great granddaughter would be about 786 years old. I don't think she would be *fine*. I think what we've learned is those boys like a *much* older woman.

Baby sharks and Brexit; acid and corals

I was doing some baby research before their imminent arrival, and I somehow stumbled across a YouTube video of a song called 'Baby Shark'. Now, if you are a parent you will probably already be aware of this song. If not, then I warn you: this earworm is going to destroy your life. I presumed it was just some random video, but then a few weeks later I found it has as many views on YouTube as there are people on the planet. Anyway, it turns out baby sharks are actually being born earlier due to warmer oceans, and they are less likely to survive as a result.[24] Although I admit 'Premature shark doo doo doo doo doo doo' has less of an appeal as a catchy kids' song.

Warming oceans also mean fish are migrating poleward. Sharks have been found in waters that were previously too cold – so my *Jaws vs Santa Claus* filmscript doesn't seem so mad now, does it, Spielberg?[25] And in the UK, our delicious cod and haddock are migrating further north. Soon, instead of traditional fish and chips, we will be having squid and chips.[26] So if you think that we've finally left the EU, then I'm afraid I have bad news for you: you're about to become French.

Then there's ocean acidification. About 30 per cent of human-emitted carbon dioxide has dissolved into the ocean.[27] This carbon dioxide reacts with the water to form carbonic acid, which lowers the ocean's pH.* Ocean acidification also lowers the concentration of carbonate ions in the sea water. These are the building blocks organisms use to build their

* It is weird that PhD has a small h, but pH is the other way round.

shells. When there is less carbonate, it takes more energy for organisms to build their shells, which might affect their development. Imagine the Teenage Mutant Ninja Turtles, except all their shells are deformed and they don't live beyond their teenage years. Acidification will seriously disrupt marine life, as some species cannot grow while others will thrive more, and this, in turn, will have knock-on consequences for fishing and those that depend upon it. It is also messing with the ability of some fish and octopuses to use chemicals to detect things around them, like mates and predators. (You don't want to get those mixed up – unless that's what you're into. I'm not going to be judgemental about a consenting act between two aquatic beings.) This acidification will take tens of thousands of years to reverse.

And that brings us on to coral reefs. With all their colours and shapes, coral reefs look kind of like – let me think of a poetic way of putting this – kind of like the great sea god Poseidon has eaten too many Skittles and puked them all over the seabed. Corals – which I have taken to thinking of as a giant conjoined half-plant, half-animal whose body is made up of thousands of colourful lips all stuck together – that form reefs in tropical waters today first appeared in the Middle Triassic Period, about 240 million years ago, but are now suffering from the warmer oceans combined with acidification. When the water is too warm, the corals expel the algae, which provide the colourful bit, so they turn white. This coral-bleaching is harming these great living reefs, which support the biodiversity of marine species (and food for lots of people). In 2016, the most famous and largest reef in the world, the Great Barrier Reef, saw a heatwave kill 30 per cent of its corals.[28] In a world 2°C warmer, there is an 87 per cent chance we will see the same kind of heatwave as in 2016.[29] Death will become the new normal for corals.

Overall, it is easy to forget about the oceans since, well, we are not sea animals. But melting ice and rising oceans are dangerous to us for a combination of reasons. Regions are dependent upon freshwater from glaciers. The ice caps help keep the planet cool. Many people live close to the sea or are dependent upon it for their livelihoods. Entire island nations will disappear. And the changes that are coming – or, in some instances, that are already here – create risk and destruction for millions.

8

Twenty-week scan

Return to your breathing

It was an unseasonably warm spring day, and together we tore open an envelope in the middle of the park. We had decided to find out there and then. We couldn't wait until we got home. The suspense was killing us.

Oh, before I continue, I should probably mention that we were now in the midst of a global pandemic. Yeah, didn't see that one coming. Really blindsided me, to be honest. Fair play. Not sure we'd have, how can I put this politely, I'm not sure we'd have 'done the business' if we'd known Corona was on the way. Needless to say, my anxiety about having a baby had been cranked up a fair few notches. Now, there are endless things I could say here on my feelings about the pandemic, but in order to keep things light, I'll say this. It is super annoying how much it has stolen climate change's limelight. I mean, everyone was just getting really into climate change. It was in the news all the time in 2019, and the UK was due to host the

UN Climate Change Conference in 2020. I mean, talk about upstaging us. How dare you!

I say all this because COVID-19 restrictions meant that I hadn't been able to attend the twenty-week scan. Instead, I had sat by myself in a park outside the hospital for about ninety minutes, while my wife donned her mask and attended the appointment by herself. I thought about her and all the other pregnant women who were having to go through these situations alone. We were both pretty sure it'd be a girl, mostly because we had lots of girls' names and barely one for a boy. And I know I'm a scientist, and it's a fifty-fifty chance, but often a feeling feels more real than a statistic (hence the existence of climate denial). So, I reminded myself it could be either sex, and that it was best to leave it to the experts to figure it out. I got bored of sitting and walked around the park about ten times. I should have brought some water. It felt like she was gone for ages. Then she texted me to say she'd be out soon.

Now, here she was. Envelope in hand.

'Let's just do it now.'

She opened it.

'BOY,' said a note scribbled by the sonographer, in what was barely legible handwriting given the gravitas of the moment.

We were having a son. A male guy. A wee chap. I think we were both a bit stunned – we pulled surprised faces at each other. Shocked. We were over the moon, but it would take a while to adjust to the idea. A boy. Our boy.

On the way home, there were a couple of lovely, awkward teenagers sitting across the aisle from us on the train talking about a computer game – or maybe a person, I couldn't quite tell. It made me think about what it must be like being a teenager nowadays. Over the last year, there had been school strikes across the world, where young people and children had taken to the streets, demanding that more be done to protect their future. The movement had been an incredible uprising. One and a half million on the streets in over a hundred countries. I wondered if perhaps our baby would be a young Swedish girl who would take a stand against the inaction of adults and injustice. It was unlikely, given that neither me nor my wife are Swedish, nor did we live in Sweden. And, as I said, we were now having a boy (that bit was taking a while to sink in). But still. I wondered what he, MY SON, would be like as a teenager. I wondered if he would go off marching with his buddies. I realise that people of my age had it so easy, growing up in a time where everything didn't seem hopeless and shit. I remember being worried about whether my parents would let me go backpacking to Venice – now teenagers are worried about whether Venice will even be there to visit. I didn't want to bring a person into a world where their life would be spent being anxious and concerned about the state of the planet.

Later that night, after the 'new dad' part of me had clearly checked out and been replaced by the climate researcher Matthew Winning, I found myself asking the question: 'Is a boy a good or bad thing for the planet?' It is not a question I'd ever thought about. I guess, on balance, it is probably worse, given that men have been in charge while the climate has gone to pot. A quick search found a study of single people in Sweden showing that men, on average, emit 16 per cent more greenhouse gases than women, although really I imagine emissions depend far more on the income of the child's parents, lifestyle, etc.

However, our son was going to be British, which historically is not notoriously great from an environmental perspective. At this point I was thinking that not only would I be having a child – apparently the worst thing for the planet – but an English male child: which the Scotsman in me struggled with.

I quickly put these stupid thoughts out of my head, brought expectant dad Matt back into the room, and watched a hypnobirthing video for the first time. For those of you less *au fait* with the term, no, it did not involve swinging a pocket watch in front of my wife's face. It simply refers to techniques to help cope with and prepare for a positive birth experience. It is essentially about how being calm helps more oxygen into the womb, which facilitates a quicker and less painful birth. It actually made a lot of sense. And so, we had started working our way through an online course that provided breathing techniques, positive affirmations and various other things that I, as a birth partner, would be in charge of, to keep things super chill.

The other news was that we had had an offer accepted on our first home. Thankfully, this had happened just before lockdown. We had been looking at properties for a few weeks, progressively getting further and further out of London in order to be able to afford anything resembling a family home and not a shipping container. Eventually, we saw a lovely place on a quaint street with a park nearby and country walks. It would make a wonderful family home. What sold it for us? There was a house along the road with a miniature model of itself in the window. The same house, but tiny, all the same details. And inside that wee house, in the window, was a miniature model of *that* house. And so on. That's when we both knew this was the street for us.

Anyway, it was now weeks later, and I was pulling my hair out most days, wondering if the move would go ahead, given that there was currently no economy to speak of, and nobody could leave their homes. In hindsight, this was perhaps not the best year to undertake ALL THE LIFE STUFF. I was beyond stressed.

Oh, and I'd looked into the numbers behind offsetting a child. I returned to that study I'd found about the climate impact of having children.[1] The central estimate for emissions associated with having a child were 58.6 tonnes a year. This was the number parroted in newspapers and articles. That is about eight times my personal emissions a year, because it's based on the emissions of your child, and your child's children, and your child's children's children, and so on and so on.* And that number is just for one parent. Needless to say, it is impossible to offset by your own actions. I calculated that, if I wanted to have a carbon-neutral baby, I would need to convince seven of my closest friends to come live in the woods with me forever, or I would need to start killing about a hundred dogs a year. Or about two hundred dogs if it's those small yappy ones. Did I have the commitment and time it would take to put this into action? That's almost a dog every second day. How would I go about it? I guess I'd need to do as many as I could in one go to make most efficient use of my time. But surely, it would be more suspicious if hundreds of local dogs went missing at once. I needed to think this through some more, maybe try to

---

* Technically, each parent is responsible for half the emissions of a child, a quarter of grandchildren, and so on for generations of descendants. Each future generation is expected to have two kids each. The total is then annualised.

write a computer code to minimise time and maximise number of dead dogs.

Or I guess I could just convince lots of people to change their energy provider?

Yes, that seemed more sensible.

9

What's with the weird weather

This is the chapter where we talk about all the disaster-movie stuff: fires, hurricanes, floods. All of these together are extreme weather – but not in the cool way that certain sports that stoners do are called 'extreme'. These are the types of climate threats that us humans understand better, because they are obviously terrifying and happen quickly. For instance, a wildfire is pretty damn scary and it feels like an immediate threat to our lives, whereas, say, sea-level rise, which is basically a cliff being eroded over the course of decades, is barely noticeable from day to day. It's a bit like the difference between cigarettes killing you over decades, or one burning your house down.

Floods

A hotter planet means more flooding. I know it sounds a bit contradictory, but warmer air leads to more evaporation, which means more moisture in the air, which means that when it rains, there's more water, which potentially means more chance of it absolutely pissing it down. Basically, when it rains, it pours. So, there is an increase in the number of days of heavy rainfall, chances of flash floods and getting absolutely sodden. For every degree the planet warms, the atmosphere can hold about 6 or 7 per cent more moisture.[1] I feel that people whose hair becomes unmanageable in high humidity have probably suffered enough already, but I'm afraid it's only going to get worse. So, quite simply, if the planet does warm by 3°C, there will be over a fifth more water in the air.*

And most of it will fall in the places that are already wet. Regional changes in precipitation trends have been identified, and there has been 5 per cent more rain and snow falling across the United States over the last century.[2] Flood disasters across Europe have doubled in the last thirty-five years.[3] Germany and Belgium suffered devastating floods during July 2021 and slower movement of rainstorms is expected to increase local exposure to intense events like this under climate change.[4] These floods were 3–19 per cent more intense and events like this up to nine times more likely because of climate change.[5] Venice is now so synonymous with flooding that I imagine the city will soon pass a law that all tourists need to take some water with them when they leave.

Monsoons are driven by differences between land and ocean temperatures. Forty per cent of the world's population live in areas that have a monsoon season; these seasons are becoming stronger.[6] In 2010, a torrential monsoon in Pakistan made 4 million people homeless. In 2020, Kenya suffered flooding that left 40,000 people displaced. The last time we went on holiday

---

* BIG TIP – invest in umbrellas.

was to Japan in 2018, where we spent a morning in a town called Atami. In 2021, Atami suffered a devastating landslide after receiving more rainfall in the first three days of July than it normally does in the whole month. Two people were killed and, at the time of writing, another twenty were missing.[7] Importantly, rainfall is also becoming more variable, meaning it is difficult to predict rain from year to year. Basically, anything dry will rise in value – crackers, Richard Ayoade's sense of humour, your mum's Christmas turkey.

The UK had the wettest February on record in 2020, recording 237 per cent of its average rainfall. It also had the wettest *day* on record, on 3 October 2020, with enough rain to fill Loch Ness.[8] I know you might not have an immediate idea about how enormous Loch Ness is, but it's big enough for us to not conclusively know that there definitely isn't a dinosaur swimming about in it. Extended periods of extreme rainfall in winter are now seven times more likely in the UK. In 2015, Storm Desmond (which was made 59 per cent more likely by climate change)[9] saw record rainfall of 13.4 inches in twenty-four hours in Cumbria.[10]

About 1.8 million people in the UK live in areas that are currently at significant risk from flooding. By the 2050s, this will increase to 2.6 million with 2°C of global warming, and 3.3 million if we were to go as high as 4°C, as sea-level rise and increased river flow from extreme rainfall combine.[11] And I have really flat feet, which means I can't comfortably wear wellies, so I'd rather that didn't happen. Glasgow, where I'm from, is expected to be one of the worst cities in Europe at risk from flooding by 2050.[12] Not a top-ten city table you want to be included in. Local band Wet Wet Wet may have to add another couple of 'Wets' to their name.

Other human factors no doubt play a significant role in flooding. People have built up towns and cities around water ecosystems, which have been disrupted. Deforestation reduces many natural flood and landslide prevention barriers, where soil would normally soak up rainfall. It can, therefore, be hard to disentangle what is climate change and what is not, because we've built up so much over the last century. What is clear is that countries will need to spend more money on flood prevention beyond what their current system is designed for – either that, or invest in a Noah's Ark. Which might seem like the easy way out, but then you've got to factor in rounding up all those bastard animals.

Flooding already costs about £1 billion a year in the UK. Also, you'd think it wouldn't need saying, but apparently it does – building new homes in high-risk flood plains might not be a smart idea. The number of new homes being built on flood plains in England has doubled in the last few years, and a tenth of new homes in England are being built in high-risk areas.[13] I cannot stress enough how highly this ranks on the list of stupid ideas.

This is not a sentence I thought I'd ever be writing, but in 2018, to combat flooding in the UK, Michael Gove released two Eurasian beavers into the Forest of Dean. I mean, he didn't just do it himself off his own back – he was Environment Secretary at the time, he didn't get on the dark web to offer a grand for a couple of live beavers. Anyway, 400 years after becoming extinct in this county, they were released back into the wild for a trial. It is hoped they will build dams and, if successful, then official adaptation policy will involve Gove running around the country's rivers, throwing beavers into them left, right and centre. Somehow, I just know that the image of Gove wearing wellies with a buck-toothed rodent under each arm will come back to haunt me in my dreams.

Ring of Fire

Wildfires have a bit of an Old Testament 'wrath of God' vibe going on. And so, they can quite easily become the poster boy for climate impacts, especially when they affect Western cities and countries. Plus, fires photograph better than floods, because they bring their own lighting.

The thing is, though, that wildfires have always existed, and in the past, there were more forests covering the planet and emissions from wildfires were higher – although deforestation has cancelled out the reduction in emissions.[14] On top of warming conditions, there are other human factors that are exacerbating the problem. Some natural burning is actually a good thing, as it can help certain ecosystems and clear debris. However, over the last century, our societies have got too good at putting out small fires, with forest-management safety campaigns being almost too effective, which leaves a build-up of lots of flammable fuel lying about – so when the big ones come, they come hard.* Basically, natural burning, which does an important job, has decreased. (This is *not* a call to arms to pyros – please don't say you're relying on the 'Dr Matt Winning defence' as the police cuff you.) So, there is more tinder than there should be. Plus, more people are moving to the nice, picturesque areas, which actually turn out to be a new circle of Hell.[15]

* Sorry, I just read that sentence back. I apologise, and hope my mother never reads this.

On top of that, the climate conditions we are creating from heatwaves, droughts and even floods, mean that these fires are now spreading quicker and further, and burning seasons are lasting longer than humans have previously experienced. Warmer temperatures are drying soils, and fire seasons are now longer across a quarter of the planet.[16] Wildfires can affect us in a variety of ways, the most obvious being burning you alive. Beyond that, there's property damage, loss to wildlife and habitats, and smoke inhalation and air pollution, which are extremely dangerous for health, especially in younger children. Wildfires can also increase flood risk, as the burning sears the ground, making it less likely to take up rainfall.

The starkest recent example of wildfires has been in Australia, which had its hottest year on record in 2019 and then suffered a bushfire at the end of the year, which burned on into 2020 and was witnessed worldwide with abject horror. A billion animals died from the ravaging fire, and images of desperate koalas and kangaroos appeared on the news and our social media feeds. These bushfires were made at least 30 per cent more likely due to climate change.[17] The smoke cloud from the bushfire, which entered the stratosphere, was three times larger than anything we've seen on earth before.[18]

Now, I struggled to see the humour in this topic. It was bleak. In fact, I had pretty much given up. That is until I found out that a company called Geeky Sex Toys were selling a 'Down Under Donation Dildo' to raise money for relief. And it was not just any plastic penis: it was gold and green to match the Aussie flag, the base of which was in the shape of Australia, and it had a wee cute koala on the shaft. Gives a new meaning to the idea of a good deed giving you a warm, fuzzy feeling inside.

In the USA, the area burned by wildfires has been increasing over the last thirty-five years.[19] Nowhere is the stark reality and contrast of a warming planet more evident than in California. The seven largest fires in the state's history, in terms of area burned, have all occurred since 2017. We have all heard about celebrities in Los Angeles being affected by these fires: in 2018, Kim Kardashian and Kanye West hired private firefighters to protect their mansion, which is a taste of the bleak inequality of things to come. That same year, the Malibu homes of Miley Cyrus and Scotland's own actor, Gerard Butler, were partially destroyed in the Woolsey Fire, which burned for thirteen days straight. If his films are anything to go on, it almost certainly happened just as he was about to retire from being a fireman for twenty years.* California will continue to experience an increase in extreme fire weather due to climate change.[20]

Pine and spruce bark beetles are infesting forests in North America, killing hundreds of trees, which are then becoming more combustible. The infestations themselves are driven by warmer temperatures, meaning the bugs are not killed off in winter. Arctic wildfires across Siberia and Alaska are becoming larger, too. These fires may not affect as many people directly, but they will darken snow and ice, which will warm the area further through the albedo effect. Plus, many occur on peatland and permafrost, which means they can release huge amounts of ancient, stored carbon and create zombie fires.

---

* Gerard is from Paisley, where I was born. If you aren't familiar with Paisley, it was up for UK City of Culture 2021 and lost to Coventry. Which is a bit like being up for an academy award and losing to Gerard Butler. Paisley is unlikely to see any wildfires in the near future. Gerard will still choose to live in California despite its habit of being on fire.

Storms

These, just like the various *X-Men* films starring the character of the same name, have been getting worse and worse. The intensity of hurricanes and typhoons is linked to sea surface temperatures, which are getting warmer.* Ocean surface temperature needs to be over 26°C for a hurricane to form, and by the time the film *Ocean's 26*, starring the reanimated corpse of George Clooney, comes out in the year 2050, that will more often be the case.

However, these storms are a trickier one to pin down when it comes to climate change. There are a number of other factors, such as wind shear, that also contribute to whether the conditions are ripe for a tropical cyclone to form. We have only really been able to track them more accurately since the 1970s, when we started using satellites.

What does the science say? Well, there aren't actually any more hurricanes than there were before. Although in 2020 we did see a record number of major storms forming in the Atlantic Ocean – twenty-nine in total (but that is just one place).[21] Maybe we will see a change in the overall trend in the future; maybe not. It is sort of beside the point. Importantly, the one thing we do know about the strongest hurricanes is that, like the acting performances of Daniel Day Lewis over his career, their intensity is increasing.[22] Warmer oceans mean hurricanes take up more

---

* Hurricanes and typhoons are essentially the same thing – they are both tropical cyclones, but have a different name depending upon where they are from. A bit like the fact that in the UK, we call them trousers, but in the USA they are called freedom legs.

water vapour, which can lead to greater levels of downpour.[23] They drink the ocean's milkshake (that's a reference for any Day Lewis fans). Therefore, those few hurricanes that *do* make land will be much more damaging. One scary element of these storms is that the destructive force of a hurricane increases substantially as wind speed does. So an increase from 75 mph to 150 mph makes a hurricane 256 times more destructive.[24]

And I like my hurricanes like I like my coffee: Category 3 or below. But the proportion of severe tropical cyclones, i.e. category 4 and 5, has increased.[25] Climate change is like steroids for them: it is making them stronger, and turning them into moody, egotistical dicks that are annoying for people to be around. Their tracking appears to be slowing down, too. This slowing increases the damage they can do, as they are able to unleash a greater deluge of rainfall from above on a certain place.

In 2017, Hurricane Harvey hit the metropolitan area of Texas and saw more rainfall than any US hurricane in history. It damaged or destroyed 300,000 buildings and half a million motor vehicles. One hundred and seven lives were lost. An article in the journal *PNAS* found that, in Houston, the rain was biblical – in the sense that its rainfall was a once in 2,000 years event, and therefore has had the chance of happening once since the New Testament.[26] (And yes, you read the acronym of that journal – the *Proceedings of the National Academy of Sciences* – correctly. Now, despite whatever smutty double entendre is going through your head, just remember that when it comes to science and the exploration of new areas, *PNAS* stands ready.) These storms cause massive economic and social destruction. Harvey was calculated to have caused $125 billion in damages, mainly from flooding.

In other parts of the world, the impacts on countries can be completely devastating – in particular, in the Bay of Bengal. In 2008, Cyclone Nargis killed 130,000 people in Myanmar as a 3-metre storm surge ended up 40 kilometres inland. In 2020, India was hit by 190 kph winds from Cyclone Amphan, causing $13 billion in damage, while Cyclone Gati dumped two years' worth of rain on Somali in just two days, leaving properties flooded and livelihoods in tatters.[27] The numbers of people living in coastal areas is growing, even in places that lie in the paths of such storms. These people will simultaneously be facing sea-level rise – which, coincidentally, is giving storm surges a higher starting point and leading to greater flood risk further inland.

Climate Cluedo

When a flood or wildfire occurs, a question that is often asked is 'Did global warming cause it?' The answer is that global warming/climate change does not *cause* events, but it makes them *more likely*. This involves a pretty new field called 'attribution analysis', where scientists calculate how global warming is tied to a single event's occurrence, i.e. has climate change left a fingerprint at the scene, or, if they're using one of those blacklights from *CSI*, is climate change's jizz all over the place? Essentially, scientists run their climate models without the effects of climate change included, and then run it with the climate change and look at the difference to see if climate change has made the probability of the event happening much higher. The World Weather Attribution group undertakes much of these types of attribution studies.[28]

An example of what we are able to ascertain from these studies is that a single event – say, a flood – would have only occurred once every 200 years if we remove the human influence element, but now, with global warming, it is likely to occur once every 20 years. Therefore, such a flood would be made ten times more likely by climate change. There will be occasions where warming has no discernible effect whatsoever on an extreme weather event occurring. However, there are occasions when we are pretty certain events *wouldn't* have happened at all without our influence: like the Siberian heatwave in 2020, which was at least 600 times more likely due to climate change, with the most likely estimate being *80,000 times* more likely.[29]

In summary, climate change is making lots of terrible, devastating events much worse – which is no good for anyone except Hollywood producers of disaster films. These events create loss and have human ramifications that last for lifetimes. I'm not sure how I can be any more succinct.

10

When will the world end?

This chapter deals with the more dramatic – though more unlikely – major events that could be triggered by climate change. Unknown levels of warming may cause abrupt changes and severe alterations in the earth's current state – like in the film *The Day After Tomorrow*, where all of a sudden the ocean stops moving and the entire planet becomes a ski resort. That sort of stuff. Except this is real and not a Dennis Quaid film.* I'll be honest, my motivation for writing this chapter is to justify some more cinematic imagery that means there might be a chance I can get on that Netflix gravy train. Here I discuss some of these critical points that, if crossed, could cause dramatic planetary shifts and create a new state of the earth from which we cannot return. And for us, a new state is not good news. It's the equivalent of Mother Earth's new look being a skinhead, with a swastika tattooed on her newly bald scalp.

---

* Although one climate solution may be to shrink ourselves down, just like Dennis Quaid did in the excellent film *Innerspace*.

Tipping points

We call these shifts 'tipping points',* a term that has had its weight somewhat undermined by the existence of the jolly UK daytime TV gameshow of the same name hosted by Ben Shepherd. To be honest, it wasn't even dramatic enough to start with: it needs to be something more like 'Next stop Fucked City, Population: us'. But then again, as soon as the phrase finally worked its way into common parlance you know what would happen: 'Hello, and welcome to the new ITV gameshow, *Next Stop Fucked City, Population: Us*, with me, your host – Ben Shepherd.'

When we think about climate impacts, we might tend to assume that double the emissions means double the harm, with things getting proportionally worse: heavier floods, worse fires, etc., as seen in the previous chapters. And while some aspects *do* have a straightforward, linear relationship, like Monica and Chandler, others are much more complicated, like Ross and Rachel, or like young people and the classic sitcom *Friends*. In fact, some aspects encounter what we call a positive feedback. Positive feedback sounds like it would be a good thing, but in this situation, that feedback would be: 'Well done, you have absolutely, irreversibly fucked mankind beyond salvation and doomed millions to untold suffering for thousands of years,' accompanied by that GIF of Captain Jean-Luc Picard sarcastically clapping.

Positive feedback effects are not positive in any way what-

---

* A phrase that entered popular culture through the book *Tipping Point* by Malcolm Gladwell.

soever. It means that the effects are self-reinforcing and exacerbate more of the same problems. They lead to exponential relationships, or even irreversible threshold events, where we quickly switch from one situation to another. Think of it like tipping over a cow. You push and push, and not much happens for a while, but eventually there comes a point where the cow falls over entirely and tumbles down into a ravine, where all the other cows find it. The herd crowds around their friend. But Daisy doesn't respond. *Why won't Daisy respond?* think the other cows. Every day, they watch their dear friend picked apart by birds. They are traumatised. They produce no milk and the farmer can't explain why. He ends up destitute. All because you had to see what happens when you push a cow.

Another analogy is to think of tipping points as being like a game of Jenga. Your choice.

Now, let's have a look at some of these positive feedbacks.

Bloody ice sheets again

Firstly, what happens if we reach a point where *all* the ice melts? Santa Claus becomes a climate refugee, sure, but is that probable? Well, the good news is that the biggest chunk of ice – East Antarctica – isn't going to go anywhere for a while. If it did melt, sea levels would rise by 65 metres.[1] However, it formed about 35 million years ago and has remained stable, even when the planet has been far hotter. Phew! The bad news is that the smaller chunks – like West Antarctica and the Greenland ice sheets – might still go bye-bye at some unknown point. These could reach a level of melting whereby they collapse completely, and this would increase the global sea level by about 6 or 7 metres for each. That's three Peter Crouches.* Higher sea levels across the entire planet would require saying adios to Bangkok, Manhattan and Shanghai, and you'd no longer need to have ironic beach bars in Shoreditch.

How likely is it that these ice sheets will collapse? Well, the honest answer is we don't really know yet. Very much worth 'keeping an eye on' then. Somewhere between 1–4°C of warming will cause this tipping point in Greenland.[2] There is a decent chance the collapse will eventually happen. It will take time, though – probably at least a thousand years or more to disappear completely (note to Netflix: this might sound slow, but we'll demonstrate with some sort of time-lapse CGI and a banging techno track). Without immediate change, we are consigning future generations to a drastically different world much closer to a nightmarish waterpark than today.[3] The amount of water that melted from the Greenland ice sheet in 2019 was enough to cover all of California in four feet of water. It didn't, thankfully.

Runs AMOC

Secondly, it turns out the main premise of *The Day After Tomorrow* (before it gets mega-unrealistic) is actually based on a real thing – a shutdown of the Atlantic Meridional Overturning Circulation (AMOC). We don't notice it, but on a day-to-day basis the ocean currents help control our global climate. There's a whole lotta water down there, transferring

* Or four Darth Vaders, if you prefer Imperial.

a whole lotta heat around the planet: the equivalent energy of a million nuclear plants.[4] This transfer keeps an equilibrium in the global climate system, ensuring an energy balance, and we have built our societies upon this stable world. (It's lucky we haven't built our societies on anything unstable, like stock markets.) However, melting ice may have the knock-on effect of slowing down and then switching off deep-ocean circulation and the Gulf Stream. Why is this an issue? Well, dense, cool, salty water travels from the Gulf of Mexico and sinks in the north Atlantic. But as more fresh water from Greenland melting mixes with the AMOC, it makes the circulating water lighter and less able to sink. This change slows down the AMOC circulation, and may eventually shut down the transfer altogether, at which point the Gulf Stream would no longer send mild air towards the European continent.[5] This shutdown would plunge Europe into much colder winters. Scotland would become much more like other parts of the world at the same latitude – like Alaska. There would be changes in tropical rainfall, and sea levels would rise by half a metre in the north Atlantic (that's on top of the sea-level rise from other climate change). Arable farming in the UK might no longer be possible.[6] Studies suggest that, already, the AMOC circulation has weakened by 15 per cent. And that is a big deal, even if you are thinking, 'That's OK, I wasn't planning on doing any arable farming.'

Fire ice

Thirdly, and terrifyingly, deep underground, there are bubbles of frozen terror. Imprisoned in both the permafrost and under the sea floor are dangerous methane clathrates. Now, I realise 'methane clathrates' sound like a second-rate *Doctor Who* villain, but to give the threat a more tangible feel, I will call them by their nickname – fire ice. Essentially, these are trapped methane bubbles from decaying organic matter that are held solid at low temperatures and under high pressure. Permafrost melting – in places where the earth is warming the fastest – and warming oceans could release the trapped fire ice, bringing these bubbles back to life and into the atmosphere to cause untold harm.

This is like the film *Demolition Man*, or *California Man*, or some other nineties film about a Man that has been frozen and is then thawed. Except that instead of Sylvester Stallone or the guy that was in *California Man*,[*] it is a potential 1,400 Gigatonnes of carbon stored in the Arctic.[7] And more methane means more warming, means more melting, means more methane. So, it's an idea that doesn't just have potential for a sequel but for a whole franchise – a franchise almost as never-ending as *The Mummy*.[†]

Now, the fire ice from oceans is unlikely to happen in the near future, but the permafrost melting is more of a distinct possibility. The permafrost stores would almost double the amount of carbon currently in the atmosphere.[8] The IPCC states that the land area covered by permafrost 3 metres deep could shrink anywhere between 37 and 81 per cent this century, depending on warming.[9] Estimates suggest that 3°C of warming

---

* I can't remember his name – but I can remember he was in the film *Airheads* with Adam Sandler and Steve Buscemi, which my brother and I rented on VHS several times in the early nineties from Blockbuster video.

† He was also in *The Mummy*. You know who I mean!

would melt sufficient permafrost to release enough carbon to warm the planet by a further 0.5°C.[10]

Forest dieback

Next is Amazon dieback. No, this is not the name of a new Bruce Willis film made by a prime streaming service. It is, in fact, the name for an event where extra warming causes an extended dry season and leads to the Amazon rainforest eventually drying out, to be replaced with a savannah.* The Amazon has over 390 billion trees and soaks up a quarter of all the CO_2 captured by trees globally.[11] So much so, that's it's referred to as 'the lungs of the world'. Well, unfortunately, it appears to be smoking two packs a day.

The concern is that a longer dry season leads to worse fires, and then those areas are replaced by dry grasslands, which store a lot less carbon, in a sort of shit climate-change gentrification. Plus, it is home to indigenous people and thousands of species. Unsurprisingly, a study by Brazilian scientists in 2021 found that the Amazon is now emitting more carbon than it can absorb.[12]

There are also similar impacts on the boreal forests, which exist at high latitudes in cold conditions across Canada and Russia, and cover an area the size of Europe.[13] These forests store carbon in the trees and in the soil. I mean, you can guess what's coming next, can't you, if you've read this far. Warmer temperatures, more wildfires, more carbon; warmer temperatures, more wildfires, more carbon, in a perpetual cycle of silly

---

* *George of the Jungle*. He was the lead in that.

buggers. It's one of our best carbon sponges, but it is getting all crappy like the sponge in your sink does after too many uses. Without it, we would lose an important place to store carbon.

None of these tipping points will necessarily happen: they are uncertain. But we know they *can* happen, and the massive risk they entail means we should absolutely make sure they do not happen. We are pushing the planet harder than it's ever been pushed before. If any combination of any of the above tipping points occurs together, we'd soon be heading towards proper disaster-movie territory where humanity cannot simply adapt. These tipping points are another main reason to avoid dangerous levels of warming, i.e. over 2°C. And the decisions we are making now and over the coming decade may well lock us in to their manifestation. Believe me, you do not want this. Disaster-movies, like *The Day After Tomorrow*, are only good if you are Dennis Quaid or Jake Gyllenhaal (rich, white men). For us, it'll be hell.*

While the world will not end, it could change unrecognizingly. If I could give one tip, it would therefore be to not let any of these tipping points happen. Otherwise it is climate Jenga for us all. (Climate Jenga is the title I'm pitching to Netflix. Though happy to take it to Amazon, or I could even see it as a new ITV daytime gameshow.)

* *Bedazzled*. With Elizabeth Hurley. Brendan Fraser! That was his name.

11
Due date

I inhale peace and exhale tension

Any. Day. Now.

I had always heard about the 'due date' and just assumed, 'Ah, that's exactly when they'll be here, then.' Turns out it doesn't quite work like that.* Nobody tells you that there is essentially a month-long window for the birth. Or that, even though the pregnancy is forty weeks, you actually start counting *before* the conception even took place. So, no matter how prepared you think you are, you've somehow lost time before you've even started.

In some ways, it had gone far too fast, yet it had also been like *Groundhog Day*. At least we had been in our new house for almost two months now. Thank goodness for that. The move had broken up the monotony, provided a change of scenery and finally allowed us to start 'nesting'. However, I will say this: moving to a new house with my thirty-two-

---

* Note: This would become a bit of a theme about having a child.

week pregnant wife during a pandemic and a heatwave had not been easy. I pretty much did all the heavy lifting as we didn't want strangers in our house. One of my good friends hired a van for us, though when he showed up, the rental company only had one van left. He arrived outside with the most ludicrously oversized vehicle, and we spent about an hour going up and down on the automatic lift while posing with aviators on, pretending we were in *Top Gun* (although I'm not sure anyone held an ironing board in that film). Also, I did manage one practical joke before we left for whoever moved into the flat after us, by writing the name of a fictitious child called Brewster on the wall at various different heights, with dates next to them, but so that it looked like he was getting smaller over time. That way, they'll think some sort of Benjamin Button person lived there. Happy to lose a part of the deposit for that.

Our new neighbours had been very welcoming, and there were a number of other young couples with whom we'd had some barbecues. Apparently, though, crime in the area was high. Cheese-based crime. We were added to the local WhatsApp group and were told the story of the infamous Dairy Bandit. One morning earlier in the year, a neighbour's doorbell had been rung by a French lady whose car was covered in cream. Her voice trembling, she asked if anyone had seen anything. Then, a week later, that same neighbour's own car was found smattered in cream too. Finally, another neighbour's car was discovered with – and I quote directly from the WhatsApp message – 'a single basic-bitch cheese slice out the wrapper' left on their windscreen. Apart from that, it is a fairly peaceful area.

We had been preparing for the birth and rewatching

the online hypnobirthing class, which was simultaneously insightful, helpful and terrifying. We were getting really into it. My wife was practising breathing techniques each night and had created a birthing playlist. I found myself muttering positive affirmations under my breath as I fell asleep at night. We had a hospital bag ready with clothes, lights and snacks. I'd stored all the relevant phone numbers for the hospital in my phone. And there was an illuminating moment when we discovered we'd been talking at cross purposes for a couple of days, after I realised the perineum was not the name of a kids' play gym.

I'd recently come back to thinking about the climate impact of having a child. I couldn't quite let it go. Did we really need to be thinking about not having children for the climate? I dug deeper into the study I'd found about the carbon footprint of having children,[1] and two points stood out.

Now, I will warn you, this part may get a bit boring and technical. I mean, not to me, I love this shit, but you'll have to bear with me. Firstly, a child's carbon footprint varies by geography, as countries have different energy systems, etc. The study had attempted to provide a number for developed countries. As such, the 58.6 tonnes quoted was the global average of three countries – Japan, Russia and the USA. The UK is most similar to Japan, which is at the lower end of the estimates, at roughly 24 tonnes. The USA brought the average right up.

Secondly, and more importantly, the future is not yet set. To quote my favourite childhood film, *Terminator 2*: 'There is no fate but what we make for ourselves.'* The number quoted

---

* I now realise it was not a particularly appropriate favourite film for a ten-year-old.

(58.6) is actually taken from a less cited study from 2009, which assumes a future where things don't get better or worse.*[2] Thing is, emissions will not remain constant indefinitely. In the UK, per-capita emissions are already about half of what they were in 2009.†

Therefore, given these two aspects, surely the number could be a lot lower depending upon where you have a child, and what happens to average emissions over the coming decades in the country your child lives in. Plus, if you raise a vegan cycling enthusiast, you're quids in.

I decided it would be best to simply contact one of the authors of the study and arrange to have a chat. Professor Kimberly Nicholas is super nice and a very thoughtful Californian woman who works at a university in Sweden.‡ It was one of her students who'd undertaken the study, and she'd co-written the paper with him.

I first asked her how she felt about the study and the attention it got. Scientists do not usually have conversations about their feelings.

She made it clear that 'Unlike driving in SUVs and flying in planes, making the free decision of whether to have children and how many you want is a fundamental human right. We tried so hard to be extremely careful with how we presented

* i.e., that per person emissions remain constant at current rates.

† However, the original countries from the study – Japan, Russia and the USA – have all been rubbish, and therefore their per-capita emissions are pretty constant. I think most would agree this will probably improve under future energy system trends.

‡ She has recently written her own excellent book about her personal experiences of what it means to work on climate change. It's called *Under the Sky We Make* and could not be more different from the aforementioned paper.

those results.' She added: 'I felt very uncomfortable with that aspect of the paper. And I mean, that was certainly not the reason that we wrote that paper, or even the main idea of it.'

The point of the paper had simply been that government schemes and school textbooks always suggest the small-impact items, like recycling and changing your light bulbs, over larger things that make more of a difference. I mean, it is hard as an academic if others take your scientific work, strip it of all its nuance and run with it – and this was a prime example of that. The problem is too complex for pithy headlines, but the media love simplifying messages.

We then chatted about her thoughts on whether people should even be contemplating such decisions about the climate when deciding whether to have a family. Was it important? Her view was, quite simply: 'Nobody is going to save the planet by not having kids. I think it's really important to realise, if you choose not to have a child, you are preventing additional emissions from being created in the future. But you're not doing anything about emissions that exist right now. And we know that the emissions that exist right now are driving us towards catastrophic climate change, [and that they] need to be cut in half by 2030.'

I hadn't even thought about that aspect.

'At current rates,' she continued, 'the entire carbon budget will be gone before today's newborns can write cursive.'

Let's put it to rest now: the climate impact of choosing to have a baby is not important. Yes, it will have an impact, but we need to solve climate change now. That's long before babies conceived today could possibly create emissions for themselves and the next generations of their descendants. The real solution to climate change lies in rapid decarbonisation and reducing overconsumption, not population. And that's fortunate because telling people whether they should have children or not is completely unethical. Population control comes with its own unique history of racism and a blatant disregard of women's reproductive rights. Buying an electric car doesn't. Climate does not trump these issues especially when other solutions exist. What I think we should take from this study is not that we should have fewer children, but that we should be aware of what every aspect of our lifestyles, including becoming parents and how we raise our children, has on the climate. Ironically, of course, we might know this already if it was included in school textbooks.

Now I felt that concern was finally done and dusted in my head, I could focus on the imminent birth of our son. In particular, we had been decorating the baby's nursery. It took us weeks to find a plasterer. In summer 2020, with less than two weeks until the due date, we were in the middle of a heatwave. For the first time since records began, the UK had just had six consecutive days over 34°C. The very polite and muscular young man who agreed to do the plastering for us came by and told us that he couldn't plaster, because it was literally too hot in the room for the plaster to dry without immediately cracking. Now, I know that is very much a 'first-world' problem. But it is not harmless. That same weather had victims that wouldn't be moving on so quickly. On exactly the same day, a train near Stonehaven in Scotland came off the tracks completely due to a landslide caused by the thunderstorms and insanely heavy rainfall that followed the heatwave.[3] Tragically, three people died.

We eventually got the plastering done when it cooled down,

and on the very day of the due date my wife was up a stepladder hanging wallpaper.* The nursery was finished just in time, and looked lovely. Everything was ready.

We awaited his arrival.

* We couldn't trust my judgement.

PART 2

CAN WE CHANGE?

12

What's causing things to heat up?

How did we get into this hot mess – and how do we get out of it?

It is all around you. The door in the room you are currently sitting in was cut down from a tree that was soaking up carbon. It was then transported in a truck, powered by petroleum drilled from the ground, to a factory that uses machines powered by electricity from a grid that burns yet more fossil fuels.* The towel hanging behind the door is made of cotton grown using fertilisers, derived from gas, and then brought here on a ship burning heavy fuel oil. The toilet you are sitting on – wait, why are you reading this book in the toilet? What is wrong with you? Please wash your hands. But washing your hands causes climate change – you need hot water, heated up by your gas-fired boiler. *Don't* wash your hands? I could go on.

Climate change is woven into the very fabric of our daily

* This depends what country you are in. The UK electricity grid is now about half low-carbon sources. So it is different in the UK than, say, Poland.

lives, like a quilt made of existential dread on sale in the Etsy store of our minds. It is all-encompassing. Each day, we all make hundreds of tiny nano-decisions that contribute to the continued warming of the planet. Choosing what to eat at a restaurant? Having a bath or a shower? Buying this book? Telling all your friends and family to buy a copy of this book? (Actually, make it two copies.)

Now, I'm not saying we should stop eating or washing, or, God forbid, buying my books. It's not really our fault. It is the system we exist in. There's plenty of stuff we as individuals have considerably less control over that, frankly, has a much larger impact. A new bypass being tarmacked. A new school being built. A forest being cut down. Climate change is a bit like *The Matrix*. It was cooler in the nineties? Well, yes, but that's not what I meant. I mean that once you become aware of the fact it's all around you all the time, then it becomes hard to stop seeing it and return to the life you once lived. Why am I using a film reference that nobody has seen for, like, twenty years? Because I'm prepping for what it's going to be like when I become a dad, obviously.

Where are we?

Annual global carbon emissions have been increasing since the Industrial Revolution, when we first started burning coal in large quantities to power the world. That is why many temperature and carbon dioxide concentration charts are shown against 'pre-industrial levels', i.e. around 1850, before we started 'all this'.

In particular, the global emissions trend has really taken off since World War Two.* Half of all CO_2 emissions have occurred in the last thirty years – since we've known how bad climate change is. A bit like discovering fat and sugar is bad for you, but sticking to your strict diet of twenty-four Cadbury's Creme Eggs a day.

In 2019, the world emitted more greenhouse gases than in any other year in human history. While the trend has been up and up, there have been a few dips in global emissions, like in 2009 due to the financial crisis, and in the mid-seventies due to Disco – everyone was too busy burning down the dancefloor back then (or maybe it was the oil crisis). The year 2020 was also a major blip in emissions because of some pandemic or other that happened – can't say I particularly noticed.

I was often asked whether COVID-19 was good for the planet. Lots of people weren't commuting, and twats on the internet posted memes saying 'Nature is Healing' next to a picture of a dolphin playing backgammon. I had to disappoint people by answering that it was only a minor blip in the grand scheme of things. The planet wasn't saved.

CO_2 emissions dipped by around 7 per cent over 2020 due to the global economy grinding to a halt for several months. But thinking COVID-19 has solved climate change is a bit like getting confused between Hannibal Lecter having stopped his killing spree and him just having a nap. He's going to wake up in a bit, and he'll get straight back to trying to wear your face.

---

* I am categorically NOT suggesting we need to have more world wars to solve climate change.

What causes climate change?

Globally, energy is the largest contributor to climate change. And, unsurprisingly, most fossil fuels are used for energy. The largest aspect of energy is the production of electricity and heat.* They're the two big players: the Messi and Ronaldo of climate change causes. But there's also transport and various industrial uses. It is possible to break the contribution to global GHG emissions down into the following terms:[1]

- **Energy use in buildings** (18 per cent): About two thirds of which is used in our homes, and the rest in offices and shops. Therefore, how we power and heat our homes and businesses is crucial – although for many of us, homes and offices are now one and the same place. I don't know whether a Batcave would come under home or business, but that would be included in this as well, and from the look of all the crap Batman has down there, his electricity bill must be huge.
- **Energy for Transportation** (16 per cent): Planes, trains and automobiles. Great film. Terrible for the planet. Road transport makes up the largest amount – almost 12 per cent. So, if you thought a possible solution to cut down on home emissions would be to put your house

---

* It is important to note that electricity is a *form of* energy. This is something people get confused about. Often, the terms are incorrectly used interchangeably. Also, electricity is often simply referred to as 'power', which adds to the stupidly confusing terminology. I therefore refer to electricity blackouts as power naps.

on wheels, then I'm afraid not. Aviation, shipping and rail make up the rest.

- **Energy for Industry** (24 per cent): Making stuff contributes a lot towards global emissions in terms of energy for manufacturing and construction. For instance, the production of iron and steel accounts for 7 per cent of global emissions.
- **Other Energy** (15 per cent): I didn't know what that was. I presumed it was things like showing my parents how to use their iPad. Things that take up all my other energy. Turns out it is unallocated fuel combustion, leakage from fossil fuel production, and energy used in agriculture and fishing.
- **Agriculture, land-use change and forestry** (18 per cent)*: Old MacDonald had a farm, and on that farm he had a significant proportion of global GHG emissions. This includes farming stuff like livestock and manure, and emissions from fertilisers on soils. But also, deforestation and crop burning.
- **Waste** (3 per cent): Landfills and wastewater, i.e. smelly bins, garbage, rubbish, trash.
- **Process and other emissions from Industry** (5 per cent): The cement and chemicals industries both have emissions that occur from chemical reactions, e.g. CO_2 is a by-product of producing clinker for cement. These

* These numbers only relate to direct emissions and thus, for agriculture, are smaller than numbers quoted elsewhere. For instance, when you take into account the whole supply chain i.e., including energy and transport, then agriculture is more like 24%. Basically different accounting methodologies give different numbers. I've kept it simple in this chapter so they all add up to 100%.

so-called 'process' emissions are mainly unavoidable and differ from using fossil fuels for energy (which chemicals and cement both also have).

Greenhouse gases

Everything above, from flying a plane to powering and heating homes to growing food to manufacturing steel results in greenhouse gases.

CO_2 accounts for about three quarters of all the impact.[2] Most of those CO_2 emissions, about 89 per cent, comes from burning fossil fuels and industry – the rest mostly comes from deforestation.

The second-largest greenhouse gas is methane – CH_4. Just over 40 per cent of methane emissions come from livestock and manure, as well as rice production, which makes up about 1 per cent of global GHG emissions. If you were thinking you could save the planet by sitting in a cave and eating plain rice, I'm sorry, but no dice. A little over 30 per cent of CH_4 emissions come from oil and gas industry leaks, and the remainder is from landfill and burning of land. While methane is less than CO_2 in terms of its total contribution to current warming, it is much more potent on a tonnage basis – it has somewhere between a twenty-eight to thirty-five times greater impact. Therefore, while some people believe we should focus on CO_2 as the largest issue, others are worried about rising CH_4 emissions and think we need to focus more on that.*

* I guess the gas is always greener on the other side. No, I am not sorry. This is what you paid for.

The remainder of greenhouse gases are made up of nitrous oxide, largely from agriculture, and F-gases, the latter of which is unfortunately not an abbreviation of a swear word. It stands for fluorinated gases, such as CFCs and PFCs, which are man-made and used in things like fridges and air-conditioning units. (Though I would say that if I was going to sum up this entire book in one word, it would be exactly that: F-gases.)

Who is emitting what?

It is not necessarily important to decide who or what should be 'blamed' for man-made emissions. More crucial is to understand why we are where we are and the key solutions to get us out of it. (Although it would make for quite a dramatic moment if I just announced, 'The person to blame is called Simon Trout and he lives at 23 Pencil Road, Letchworth, LT4 2HE. Find him and kill him.') But there are some tricky discussions around what and who is causing emissions.

Which country is the most responsible? Well, climate change sort of makes the very idea of countries seem like a pretty irrelevant concept, given that it does not particularly give a flying hoot where emissions originate from. However, at present, the answer is China (China, China, China), which was responsible for around 28 per cent of all CO_2 emissions in 2019.[3] I'm sure they won't mind me saying this – as it's notoriously easy-going about being criticised. China was expected to overtake the USA as the largest emitter in around 2025. In fact, it happened in 2006, after China's emissions grew at a staggering rate of almost 10 per cent a year in the 2000s. The fate of the world's temperature trajectory hangs to a big extent in Chinese hands.

However, while it may well be controversial, the political system in China does allow for remarkably swift transitions if deemed to be in the national interest. Therefore, the Chinese version of communism could well play a crucial role that either saves or dooms our collective future.

The second-largest annual contributor is the USA, with about 13 per cent of total CO_2 emissions worldwide. Knowing Americans, they'll probably be livid that they've got the silver medal. However, if you start counting emissions from the beginning of the Industrial Revolution, then the USA is the single most responsible nation, with around 25 per cent of all cumulative historical CO_2 emissions. Therefore, a quarter of all global warming to date is due to the actions of the USA – around 0.25°C. In fact, China is only responsible for about half the level of historical emissions as the USA.[4] So, there you go, chaps – there's your gold medal.

The EU currently has the third largest annual CO_2 emissions at about 8 per cent of the global total, if you consider it to be a single region, which is how it is viewed under the Paris Agreement. Historically, the EU is not far behind the US with 22 per cent of cumulative CO_2 emissions.

The UK is not included in the EU numbers and sits at about seventeenth on the list, producing about 1 per cent of global greenhouse gas emissions.[5] Yet the UK is responsible for around 5 per cent of all historical CO_2 emissions. God save our gaseous Queen.

But there is another complication. These figures are based on where gases are emitted, which seems pretty sensible, right? Well, that depends on who you believe has more responsibility and power to act, the producers or the consumers. Because international trade means these are often in different countries. Made in China. Consumed in the UK. The debate over production- versus consumption-based accounting methods for emissions is one of the big rivalries that's been raging in the climate-academia world. It gets so tense that sometimes they simply finish their emails with a 'Thanks' rather than a 'Kind regards'. Absolutely savage.

The production-based approach, which I provided above, is the one used by the United Nations conventions. However, other methods looking at where things are consumed unsurprisingly often show that Western nations import more emissions than they export.[6] Ultimately, both the producer and consumer benefit from the trade, so it would probably be sensible to jointly share the emissions portion. But that complicates things, and besides, nobody likes sharing, otherwise tapas would be more popular. Border carbon adjustment mechanisms (BCAMs) are a tool that could shift responsibility by taxing the carbon content of goods that are imported and exported – the EU is introducing one soon.

Another way of looking at things is through the lens of emissions per capita. If we divide the total emissions of a country by the number of people in that country, the contribution per person looks very different. At a global level, the world average CO_2 emissions per person is about 4.7 tonnes.[7] In 2019, the USA emitted 16 tonnes of CO_2 per person and Australia 16 tonnes of CO_2 per person, compared to 7.05 tonnes in China and only 1.96 tonnes in India. Other big per capita emitters under this methodology include oil producers such as Qatar and Saudi Arabia. Consider then that a person in the US has on average over eight times the emissions of the average Indian and over double that of the average Chinese citizen.

Now, while we are talking about population, one question

I often hear is: 'Isn't the problem that there's just too many people?' Overpopulation is a tricky issue, and is far more nuanced than this oversimplified 'hot take' (which I often get from ageing bald men, often with racial undertones). The truth is that global population is growing – absolutely. We are likely to reach almost 10 billion people by 2050.[8] However, we know that not all people emit the same. And most of the population growth will occur in countries with low per capita emissions. An extra person in the UK would create 60 times the amount of emissions as an extra person in Rwanda.[9] The problem is overconsumption in developed countries.

The average UK citizen is responsible for about 5.5 tonnes of CO_2, and when you add in other greenhouse gases, it comes to a total of about 7 tonnes. That is an average of 7,000kg of climate pollution that each and every person sends up into the atmosphere, each and every year. I know it feels like emitting over 7,000kg of a gas would be something you'd remember doing, but trust me, it's happening. To put that in context, you are allowed to check in a 20kg suitcase for a flight. That is 350 suitcases that we are each pumping out there without much of a second thought.*

* And if you were to take those on a Ryanair flight, it would cost you somewhere between £7,346.50 and £13,996.50 in excess baggage.

13

Do fossil fuels suck – and can't we just suck them up?

Before we crack on with how to solve the climate crisis, let us first have some background on the main cause of the problem: dem fossil fuels. Coal, gas and oil – which was the original name of the band Earth, Wind and Fire – these still dominate our global energy system. Basically, dead plants and dinosaurs are coming back to haunt us and get their revenge. Remember how scary it was in *Jurassic Park* when the velociraptors learned to open doors? Well, it turns out they were playing the long game. Clever girls!*

* OK, technically it is only dead plants and not really dinosaurs. The comedian in me allows the artistic license for a good joke but the stuffy academic in me requires this footnote to clarify the truth. You are watching my different personalities being pulled apart like a tedious Jekyll and Hyde.

Fossil Fuels

Coal

Coal kick-started the Industrial Revolution. Burning these black carbon-rich lumps powered James Watt's steam engine, produced iron for ship-building, and created the smelly factories in England that William Blake, in the well-known hymn 'Jerusalem', aptly called 'dark Satanic mills'. Since then, coal has powered humanity towards the current pinnacle of human civilisation, with its raised living standards, life expectancy and blood pressure (from reading Twitter 24/7). But, just like building a hotel on a Native American burial ground was a bad idea in *The Shining*, building our entire modern society on the squished-up remnants of prehistoric leaves has unleashed a curse.

Coal is the most polluting of the fossil fuels in terms of CO_2. For instance, a gas power plant emits around 50–60 per cent less CO_2 than a coal plant to create the same amount of electricity. Around two-thirds of coal is used for electricity generation, and coal is also used to create the high temperatures needed for industrial processes such as iron and steel production.

Its production worldwide has almost doubled since I was born.* In 2019, coal still provided about the same contribution towards the world's electricity supply as all the low-carbon alternatives combined.[1]

I've had negative associations with coal since I found out at a very early age that Santa Claus only gave a lump of the black stuff to children if they'd been naughty. That seemed

---

* A similar increase to the amount of people saying 'could of' when they mean 'could have'.

unnecessarily cruel – just give them nothing. I therefore cheered during the 1980s, as a toddler, when coal mines were shut down by Margaret Thatcher, as I thought this would limit Santa's supply and drive up prices too high for him to compete, thereby limiting the chances of me getting a lump for Xmas (I was an economist from a very young age). Or maybe Santa has secretly been propping up an ailing coal industry for years? Anyway, worldwide, coal consumption has been on an upward trajectory for most of the past few centuries, although, like my work's weekly five-a-side football team performances, it appears to have peaked and slightly declined over the last few years.

Two countries in which coal consumption has declined are the United Kingdom and the United States (two places which I would very much not describe at present as 'united'). At its peak in the 1920s, 1.2 million people were employed to mine coal in the UK. In 1965, coal made up almost 60 per cent of the UK's energy supply. However, there was a significant switch towards gas throughout the 1980s and 1990s, which reduced coal's role somewhat. More recently, wind power and other factors have wiped it out almost entirely: coal only provided 3 per cent of the UK's energy in 2019.[2] In many ways, the UK has acted as the canary in the coal mine when it comes to coal mining.

Donald Trump often made references to the idea of 'clean coal' as a way of boosting US production again, which is like describing the chips and cheese you purchased at 3am after a wonderful evening partaking in some festivities in the Garage nightclub in Glasgow as a 'healthy takeaway' because you refused the offer to add mayonnaise.

It actually doesn't make much sense going forward to use coal for electricity, especially for developing countries. Firstly, it is becoming a comparatively more expensive energy source in relation to alternatives, such as wind and solar. Even though we have got pretty good at digging it up, it still takes lots of time and effort, plus the efficiency of global installed capacity of coal plants is not improving much. And secondly, the largest coal reserves are in North America, Russia, China, Australia, India and others, so many developing countries would therefore be dependent upon importing this coal.

Gas

Natural gas is a stupid name and perhaps the best branding initiative in the history of marketing. If I see the word 'natural' in front of something, I am usually expecting a bottle of overpriced essential jojoba bodywash from a health store, and not a gaseous non-renewable hydrocarbon that you set on fire. This makes it sound like a nice, safe fluffy cloud you can trust. It has been suggested that 'fossil gas' would be a much more apt name, so that's what I'll call it. Compared to coal, fossil gas is often seen as the lesser of two evils because it emits less CO_2 when burned.[3] Like a strangler who's bothered to warm his hands first. However, the extraction of fossil gas, its infrastructure and transportation also lead to methane leakage, and, as I mentioned before, methane is a powerful bitch.[4]

Fossil gas got big in the nineties (just like the novelty American rock band Smash Mouth), and yet thirty years later, it is still kicking around doing untold damage to people and the planet (just like the novelty American rock band Smash Mouth). Fossil gas is used worldwide for electricity production, industrial heating processes, heating homes and cooking/barbecues. The largest producer globally is the USA.[5] The global consumption of fossil gas has shown no sign of abating, increasing by 70 per cent from 1990 to 2018.[6] The fact that fossil gas is easy to store previously gave it a good advantage over electricity when the two went head-to-head to be the energy source for things like home heating and cooking, but this may well shift soon as we get better at storage and batteries.

The ability to reach more reserves of so-called 'unconventional' gas through the process of hydraulic fracturing for shale gas recently led to a gas explosion – in both senses. It resulted in a 'fracking' boom during the last decade, which saw cheap energy for the USA. However, it has not taken off to the same extent elsewhere. Unfortunately for the fracking industry, they weren't so blessed on the name front – compared to 'natural' gas, fracking sounds more like a lewd sex act. Plus, fracking has a bit of a reputation for causing earthquakes, and releasing chemicals into local water supplies. Where it can be done far away from people, it can be an effective way to access harder-to-reach, and previously too-costly reserves, one reason it has had significant uptake in the vast United States. But fracking is unlikely to take off at all in densely populated areas, as the UK and Scottish government have admitted.[7]

Oil

We use oil for driving our cars, for making plastic, for creating shit environmental protest art, and for cleaning off seabirds in a montage set to sad music. And that hasn't really changed yet. Also known as crude oil, petroleum or black car juice (or 'gas', if you are American, even though it is not a gas, it's a liquid, you complete lunatics), oil is seen as a crucial commodity on which much of our economic system relies – as demonstrated when the global economy crashes from oil-price shocks. It seems slightly ridiculous to have so much of our well-being tied up in a substance made of dead algae and plankton that takes thousands of years to form – and that's before you then have to queue for it at a petrol station. The cherry on the cake is that we even have a famous global cartel called OPEC that's effectively headed by a totalitarian monarchy which essentially controls the global oil supply. That seems like a sensible way to run the world.

To quench our thirst for oil there has been exploration into more and more remote and difficult territory (e.g. deepwater, oil sands, Arctic, more fracking, etc.). One of Donald Trump's last acts before leaving office in 2021 was an auction on drilling rights in the Alaskan Arctic. Thankfully, it went terribly.[8] As well as its horrific events, such as the Exxon Valdez spill and the Deepwater Horizon disaster, the oil industry and working on rigs are often depicted as manly, patriotic work, and thus have provided the setting for some terrible movies. For instance, in the 1990s there was a film starring Bruce Willis as the lead singer of Aerosmith who also worked on an oil rig – the film was called *Ar-ma-jay-don*, and I believe it was French. In *Ar-ma-jay-don*, there is an asteroid the size of Texas that is going to hit the earth, so they send Bruce Willis, Ben Affleck and a bunch of oil miners up to drill it – rather than training astronauts to drill, which is clearly the easier option. The irony of their day jobs is completely lost on everyone.

'Who should we get to save the world?'

'I know, let's ask those guys who are slowly killing the planet to use their skills to *save* the planet so they can get back to destroying the planet.'

Capturing carbon

Will the coal, gas and oil industries find a work-around to keep their businesses going? Well, there are also a few technological options for trying to stop so many emissions from fossil fuels while being able to keep burning them. One option is carbon capture and storage (CCS). CCS simply refers to a process whereby you collect the carbon dioxide right from the polluting process where it is being emitted and then store it somewhere for safe keeping – underground, underwater, pickled in a jar. It is the liposuction of climate change solutions. A tummy-tuck technology fix. You can keep your coal plant open, you just need to spend a small fortune on extra machinery that traps all the bad emissions. Sounds like a pretty obvious solution for wealthy companies with lots of fossil reserves to burn. Being able to retrofit CCS on to existing plants may well be necessary, as otherwise they'll keep pumping out dirty emissions. It also may be a good way to get buy-in from communities reliant on fossil jobs.

However, thus far, there has been close to bugger-all progress on this. Sorry to keep using such scientific terminology. According to the IEA (the International Energy Agency), only 40 million tonnes of CO_2 were captured in 2020, about 0.1 per cent of energy-related CO_2 emissions that year, and there are only twenty-one facilities fitted with CCS operating worldwide.[9] I think there are more branches of Pret A Manger within a 2-mile radius of my office.

It is therefore worrying that many of the climate mitigation scenarios run by scientists such as myself assume significant reliance on CCS in the future, despite the fact it hasn't really taken off yet. The IEA's net-zero scenario for 2050 requires 190 times the current amount of captured CO_2.* Why the fossil industry has decided not to invest much so far in a technology that would prolong their business indefinitely, and instead chooses to spend their money on seeking out more and more new reserves (and

* This includes both Biomass energy with CCS (BECCS) and Direct Air Capture (DAC) too. Excluding these the answer is 134 times. God, I love being a nerdy scientist sometimes.

paying mountains of cash to their shareholders) is as much your guess as mine. A big reason may be that there has not been a strong enough price on carbon to incentivise them. Perhaps they are betting that the taxpayer will eventually fund it all.

Beyond the world of fossil fuels, CCS is going to be an essential technology for the cement industry which causes unavoidable CO_2 emissions in its production, and it is unlikely we will be needing less cement, especially for new infrastructure in the developing world. Therefore, it's clear CCS will play some role in our collective future.

According to a 2021 report on net zero from the IEA, we are required to stop any exploration of new fossil fuels if we want to achieve a net-zero energy system by 2050.[10] However, fossil fuels are still supported across the world by government subsidies, i.e. people's money. The International Monetary Fund (IMF) estimated that global subsidies, including the unpaid costs of environmental damage, were $5.2 trillion in 2017 – that's 6.5 per cent of global GDP propping up polluting industry.[11] Importantly, however, there are lots of jobs in the coal, oil and gas industries that need to be transitioned to other cleaner long-term careers. This may be the hardest part of all in the low-carbon transition – making sure workers are not left behind. Governments across the world need to be extremely aware of not repeating the same mistakes as in the past, where industries were shut down without retraining or plans in place for how to support those local communities. The worst thing has been claims by those such as Trump about coal jobs coming back, which is simply a lie. We need a transition plan, not false promises.

I am very thankful for the Industrial Revolution and everything it has brought us. Well, most things. I could do without colonial oppression and fidget spinners, but you catch my drift. The

Industrial Revolution, driven by burning fossil fuels, has provided us many societal advancements and lifted millions out of poverty. Now, however, is the time to Marie Kondo those fossil fuels. We should be grateful and thankful, give them a kiss like Marie Kondo would do, and then simply leave them in the ground.

Greenhouse gas removals and geoengineering

Most of the ways to stop greenhouse gas emissions easily fall into two camps. You either swap dirty fuels and practices for clean ones or you simply do less of the dirty activity.* I was trying to think of a snappier way to express this. Swap it or stop it? Clean up or piss off? Anyway, in essence, it boils down to either using less fuels or using cleaner ones.

However, there is one other option. One that would allow us to have our dirty cake and eat it – greenhouse gas removals. While this sounds like something you'd see on the side of a big van, it actually refers to a number of ways to trap harmful gases from the atmosphere and then store them – like they do with ghosts in one of the best films of all time, *Ghostbusters*.†

---

* Or, as is often the answer, you do a combination of both.

† Except watching it back now, you realise that the Environmental Protection Agency are the bad guys in the film, because it is the 1980s, and so any sort of regulation is viewed as stifling against free market entrepreneurs. We are supposed to side with the guys who have unlicensed nuclear accelerators on their backs, and be angry that someone is trying to check this won't harm the public. And as well as attacking the EPA, they also smoke a hell of a lot in *Ghostbusters*. Make of that what you will when you come to read the chapter about climate denial and the tobacco industry. And I was really looking forward to watching it with my child.

These greenhouse gas removals can be through natural processes or human technologies. Natural processes here refers to the various elements of the natural world that provide *carbon sinks*, such as plants, soil and oceans. These can be expanded to some extent such as by replanting lost forests or restoring mangroves. Human technologies include things like direct air capture, where machines suck CO_2 directly out of the atmosphere (this is currently only done on a tiny scale).

We're flooding the downstairs flat

Let me explain with a bathtub analogy.* Think of the atmosphere as a bathtub, where the tap is the sources of emissions, water is the emissions and a plughole is the natural sinks. Imagine me lying in the bath with a mojito and a rubber duck. In this analogy, I represent all of humanity. Over the history of me being in the bath, the water level has gone up and down a wee bit as the tap runs and the plughole opens and closes, but it's always lovely and I can still breathe fine.

But recently, the tap has been opened wider and wider, so the rate at which the bath is filling up is increasing and the drainage can't keep up. The water is now lapping at my face; I've got bubbles in my mouth, and if I don't do anything soon, I'll be found by a neighbour, floating naked with mojito in hand.

What can I do? I can turn the taps down or off (i.e. use fewer fossil fuels). Yes, good idea. And I can enhance drainage (i.e. natural sinks) by opening the plughole completely. But

---

* Climate scientists love a bathtub analogy. Also, I realise that using a bath to explain a sink sounds like I really don't understand plumbing, but bear with me.

if the rate at which the bath is filling up is quicker than the drain is emptying, then the water level is going to keep rising and I'm a goner. Most typical baths have one plug – let's say this represents the full potential of natural carbon sinks. These only have a certain amount they can drain at a time. And, actually, there's less drainage than there should be, due to loads of hair clogging up the plughole (deforestation!). We could unclog the drains and try to widen the plughole a bit with a chisel (planting trees). Another option is to add another plughole – these would be the technological sinks (Direct Air Capture). But they cost a lot, because you need to call a plumber and put in another extra drainage system.

On this issue of decarbonisation compared to using carbon sinks, Dr Jonathan Foley, the executive director of Project Drawdown, put it to me well and far more succinctly: 'I try to focus on the balance of sources and sinks. Reducing sources is the priority. Maintain and enhance sinks as a second line of solutions. If your bathtub is overflowing, destroying your house, you turn off the faucet before you grab a mop.'

Back to nature

Retaining and expanding these natural GHG sinks can play an important but limited role. The best option is planting trees, which is often touted as a fantastic solution that will save us all. And true, we absolutely should have more forests anyway – for various other reasons, such as supporting biodiversity and having places to shoot Scandinavian detective dramas.*

* The only downside I've found from moving to the countryside is that, due to years of Netflix bingeing, I expect to find a body whenever I go for a walk.

However, we would need to plant a hell of a lot of trees to absorb all the carbon we're now emitting. We have the space to add more on top of the current forests we already have. Best estimates suggest the upper potential would capture about 100 gigatonnes of carbon – about ten years' worth of current emissions.[12] Even if we managed to grow them all, what if they all burned down somehow – I don't know, say in one of these wildfires we've been having recently? Then we would be, and I don't say this lightly, fucked beyond belief. We can barely protect the forests we already have. Maybe start there. Get Liam Neeson's character from *Taken* involved. 'If you destroy these trees, I will find you and I will kill you.' Deforestation already contributes significantly to climate change, so stopping that should be a top priority. So, absolutely, yes to more trees. But if you think tree planting is 'the most effective way to tackle climate change'[13] then you are *bark*ing up the wrong tree. It doesn't tackle the *root* of the problem, and we need to *leaf* fossil fuels in the ground, etc., etc.

Restoring soils is another important way to boost natural carbon sinks. Good old mud. Peatlands and wetlands, which have soils with very high organic content, also play an important role.* In most ecosystems, plants take in carbon throughout their lifetime, then die and decompose, and it all goes back into the atmosphere. But peatlands form in very rainy areas where the land is always wet. You can usually spot a peatland scientist by their enthusiasm for really good wellies. Microorganisms that manage decomposition aren't happy and don't work well, so

* Therefore, benign, sweet old ladies who put peat compost on their allotments are doing more to destroy the planet than they could possibly imagine.

instead of all that stored carbon going back into the atmosphere, it stays in the ground.

Peatlands may not be the most effective ways of absorbing carbon – they also produce a fair bit of methane – but they already hold incredible amounts of carbon: approximately 600 gigatonnes, in fact, similar to the amount in all the world's forests.[14] So it's perhaps more important to stop damaged peatlands from becoming a source of emissions as the climate warms and the peatlands dry out. Currently, damaged peatlands cause about 5 per cent of all anthropogenic CO_2 emissions.[15]

Some people have even claimed soil can completely solve climate change. In his 2013 TED Talk, biologist Allan Savory describes how holistic grazing – i.e., getting free-range cattle to roam, graze and shit across semi-arid lands – can improve soil health and solve climate change. While this may or may not work to absorb some emissions, it cannot happen at anywhere near the scale required to *solve* climate change, as he suggests. There is totally a joke about how his surname is savory and he's trying to get rid of deserts. But I cannot for the life of me be bothered putting the effort into working it out. Sometimes, you just need to know when it's not worth the hassle.

Mad scientists

There are also several technological options for sucking up carbon from the air. One is to combine bioenergy with CCS, a technology known as BECCS – which is pronounced like the nickname of a posh young woman called Rebecca that I probably went to university with. The idea goes like this: trees are planted, suck up a bunch of CO_2 from the atmosphere, then are burned for energy with the emitted CO_2 captured and stored underground, same as for normal CCS. The whole process is then repeated with new trees being grown on the same land, burned, and their emissions stored. Over time, the process would draw down carbon from the atmosphere and store it underground.

The problem is that bioenergy competes with land available for food production and so can increase food prices.[16] Another option in a similar vein to CCS is Direct Air Capture: this is literally sucking carbon out of the air, like a dehumidifier does with water. The problem is that it's also pretty tricky and expensive to do this.[17]

Then there are some even more extreme suggestions about how to manage our global climate system that sound like something straight from a *Bond* baddie playbook – these are often described as 'geoengineering' techniques. One of these ideas is to refreeze the Arctic, though I'm a bit wary whenever we think we can solve our problems by treating the planet in the same way we would a Vienetta. Another suggestion is to reduce the amount of sunlight reaching the earth by putting giant mirrors in space. Yip, like full-length ones.

Geoengineering is the plot of the magnificently bonkers Korean film, and subsequent TV series, *Snowpiercer*. In the movie, lots of light-reflecting gases are poured into the atmosphere to try to stop climate change. It backfires and creates a snowball earth. Then everyone left alive on Earth lives on board a train.* Anyway, Harvard scientists are undertaking an experimental project that tests exactly this concept

---

* If you haven't seen it, it is cray-cray – and that's before you end up watching endless videos about the similarities to *Willy Wonka & the Chocolate Factory* on YouTube.

of stratospheric aerosol release to cool the planet.[18] Get your tickets ready. Also, I presume geoengineering is where the name for the Gerard Butler blockbuster *Geostorm* came from. If you haven't seen *Geostorm*, then, briefly, the plot involves Butler having designed a satellite system that controls the climate in order to stop climate change. He's a hero. But now that system has been hijacked by terrorists, and the only person who can stop it is the now exiled maverick that is Gerry Butler. Go and watch this and then *Snowpiercer*, and then decide whether you think geoengineering is a sensible strategy for tackling climate change (or writing a film about).

Removing greenhouse gases from the atmosphere may well be essential, but it is always secondary to leaving those fuels exactly where they were and not burning them. As you will see in the next chapter, we need alternatives.

14

What about when the wind stops blowing or the sun stops shining?

None of us understand how electricity works. I mean, I think I do, but then I don't. It's as close as we will ever get to actual magic. Sure, we get that the Emperor from *Star Wars* cannot *actually* shoot electricity out of his fingertips, but he might as well be able to for all I really comprehend. My old mate Ian is an electrician – he is the real magician. As far as I am concerned, he and his colleagues are the real Paul Daniels and David Blaines of the world.

Electricity is currently (pun intended) a massive cause of climate change, but it doesn't have to be. We can use less electricity – which seems unlikely – or we can use alternative methods beyond fossil fuels to generate all our electricity. Also, it is vital that we make electricity clean – as then electrification can solve lots of our other dirty stuff, such as our cars and our heating. Electrify the world. Power to the people!

There are a variety of ways to generate low-carbon electricity

– some more established than others – and these can either be *renewable*, such as wind and solar, or *non-renewable*, such as nuclear or some types of waste (personally, I'd consider burning your own excrement to be renewable). Overall, all these power sources are low-carbon in the sense that, while there may be some emissions involved in their production, they are considerably lower than fossil alternatives. And will become even lower as we decarbonise the products that make them – like low-carbon steel production for wind turbines.

There is a perennial misconception about renewables that because it takes energy to make them means they are no use. It's simply not true. For instance, the average wind turbine has a lifetime of about twenty to thirty years, and recent reports suggest it only takes between one and two years to offset the emissions associated with making and operating it, sometimes much less.[1] In China, generating 1 kilowatt-hour of electricity from wind creates only 4 per cent of the emissions of generating the same amount from coal power.[2] Again, hundreds of scientists have been undertaking these types of studies, called Life Cycle Assessments, in order to check that low-carbon electricity is exactly that. It's always fun meeting people who think we haven't thought these things through or checked them. I wonder if this is what it's like for lots of other professions, too. I suppose people often say to their doctors, 'Ah, but doing more exercise means I will be hungrier, and so I'm not sure that's a good idea?'

The great clean energy transition has already begun. It is estimated that wind and solar combined will overtake coal as the largest source of electricity generation by 2024.[3] Many countries are already on rapidly changing electricity paths. In the UK, 75 per cent of our emissions reductions since 2012 have come from the power sector. In 2020, 97 per cent of Scotland's electricity came from renewables[4] and, for the first time, more of the UK's power came from renewable sources than from coal and gas combined.[5] However, to meet our energy demand, we will probably have to double the UK's electricity supply by the middle of the century. If this increase is to come from low-carbon sources, we'll need four times as much of that.[6] We're going to have to reintroduce national service for anyone over the age of ten with at least five years' experience with Lego.

Here comes the sun

It's mad to think that when I was writing '80085' on a hand-held calculator at school, I was using solar power and never even noticed it. It makes me think about what other grown-up things we had versions of during our education. For instance, sharpening your pencil at the bin was the childhood equivalent of taking a cigarette break.

Solar power has exploded over the last decade as it has become more affordable – costs plummeted by 89 per cent.[7] Imagine if, in ten years' time, a pint of beer went from £5 to 55 pence. You'd be ecstatic. You'd be at the pub every night. Between 2010 and 2020, the amount of solar energy capacity in the world increased 18-fold.[8] In a 2020 report, the normally over-cautious International Energy Agency (IEA) called solar 'the new king of electricity'. The potential is massive. It has been estimated that a 600-mile-square section of the Sahara Desert would produce enough solar energy to power the entire world.[9]

There are a few types of solar power available. The main

technology is photovoltaic (PV) panels, which you'll have seen on people's roofs. They look like a Doc Brown invention from *Back to the Future*. The largest PV array in the world is the Bhadla Solar Park in India, which opened in 2020 and has a capacity of 2.2GW. In *Back to the Future*, the DeLorean needs 1.21GW of energy to travel back in time. So the Bhadla Solar Park could power about two Marty McFlys at the same time, hundreds of thousands of homes – or, if you're old school, then it's the equivalent of about 2.9 million horses.[10]

Obviously, solar panels are also pretty handy for those who want control of their own power supply. One of the great advantages of solar is that it can help provide access to electricity in rural places, for example in Sub-Saharan Africa, where an estimated 600 million people have no grid connection but an abundance of sunshine. We're not talking trendy, off-grid-by-choice minimalist lifestyle hippies, but people in places like Kenya and Zambia, who would love to have electricity but cannot. Companies such as Azuri Technologies, Namene Solar and BBoxx, as well as charities like Solar Aid and projects such as Lighting Africa, have provided electricity to literally millions of people. Solar power could be a real leveller, since the technology's getting cheaper and nobody owns the sun. Well, not yet. It is surely the next step for these billionaire LinkedIn lunatics, such as Bezos and Musk.

There's also concentrated solar power (CSP) systems, which are a massive array of mirrors, often in an enormous field. This is not because the operators are extremely vain, but because the mirrors converge all the sunlight into a central spot and this is converted to heat, which drives a steam turbine.

The answer is blowing in the wind

While the UK's solar potential is limited due to the grey weather, being an island does mean we have plenty of potential for the other main renewable energy source – wind power. On Boxing Day 2020, wind power accounted for over 50 per cent of Great Britain's electricity – a new record.[11] It is likely I will look back at this comment in a decade and it will seem utterly trivial. Prime Minister Boris Johnson has said he wants the UK to be the Saudi Arabia of wind power.[12] Which got me wondering what Saudi Arabia is the UK of? Perhaps it's the UK of 'not drinking'. The UK has therefore committed to 40GW of offshore wind by 2030, which would power all of the UK's homes. It makes sense for the UK to go big on wind, because the time when we use the most of our energy – in winter – is also the time it is windiest. The same way that it makes sense for warm regions to use solar when they need it for cooling during the sunniest months.

One big winner from all the offshore wind is Her Majesty the Queen, as she owns the seabed and will receive an estimated £220 million a year from licences.[13] So I guess if solar is king, then wind is the queen. I hope she uses it to properly insulate Balmoral and sort out some low-carbon heating. Must be a bugger to keep warm. Or even better, she could move into a small bungalow. Good to know that it's still the wealthy landed gentry profiting from renewables in the UK, rather than the money being put into a green sovereign wealth fund, like Norway successfully did for oil, or, God forbid, to help local communities. Some things never fucking change. Everyone is getting involved now. BP have entered the wind market, buying some of these sites the Queen is auctioning.[14]

As a Scotsman, I know a thing or two about wind. I was born in a gale (that's my mum's name). Scotland's electricity is already dominated by wind, which makes sense. Like those Inuit tribes that have hundreds of words for snow, us Scots have many different words for wind. In fact, in the first half of 2019, Scotland produced enough wind to power two Scotlands.[15] (Nobody needs two Scotlands. What would we do with them? Would we keep one to be the 'good' Scotland we put out when guests come round? Plus, I am not sure that 'a Scotland' will take off as a unit of measurement.)

Scotland's Beatrice windfarm, which became operational in 2019, has eighty-four turbines, all with 75-metre blades (that's thirty-five Peter Crouches), and is capable of powering 450,000 homes.[16] And the Moray East windfarm came online in 2021 and has just beaten it by 20MW to become Scotland's largest.[17] An even bigger farm, Seagreen, which could power 1.6 million homes, is under construction.

The physical size of offshore turbines is now massive. The Dogger Bank windfarm, which is being built off the coast of Yorkshire and set to become the world's largest, will use 107m-long blades in 600 turbines almost the size of the Eiffel Tower. We just need to work out how to charge people to go up them and surround them with expensive rip-off cafés.

Wind turbines have been the subject of some ire, though. People such as Donald Trump and Vladimir Putin have suggested wind turbines are bad as they kill birds. Which is true, especially if badly placed in migration paths, with the most at-risk being raptors, like eagles. Wind turbines probably kill between 10,000 and 100,000 birds a year in the UK. However, cats kill an estimated 55 million birds a year.[18] Should we get rid of all the cats as well? Well, yes, absolutely. Screw cats. They should all be incarcerated on a sort of cat death row. But, just a reminder that fossil fuels kill millions of *people* a year. And while habitat assessments should absolutely be undertaken for all kinds of energy, including new wind turbines, the fossil fuels being replaced will definitely kill more birds than wind turbines do through things like mining, onsite collision and poisoning.[19]

Renewables obviously have the advantage of being exactly that – renewable. We won't run out of wind and sunshine – until the day when the sun triumphantly explodes and we all die in a gigantic fireball of molten light. Plus, renewables do not need cooling water, unlike thirsty thermal energy plants. And with climate change availability of cooling water may become an issue in certain places.[20]

However, these low-carbon options are not without their challenges. The question about reliability is a fair one, i.e. what happens when the sun doesn't shine, or when the wind isn't blowing? These are crucial questions for meeting times of peak demand, as we all tend to switch on our kettles at the same time of day. Luckily, people actually check this and don't just randomly plonk down solar panels and turbines in any old place. There are literally thousands of scientists and companies worldwide working on addressing these issues. A recent study from the University of California, Berkeley, has shown that the USA can dependably meet 90 per cent of its electricity needs from renewables by 2035 using current technologies and without increasing consumer costs.[21]

In February 2021, wind power was falsely blamed for causing massive power outages in Texas when a polar vortex winter storm meandered off course and caused four million people to lose electricity. Fox News and Fox Business blamed renewables 128 times in two days,[22] although there were actually double the outages from fossil fuel and nuclear plants than for wind farms.[23] The real reason for the outage came down to poor preparation to make the grid and infrastructure ready for such events

Storage space

Because of the variable nature of renewables, flexible electricity grids are going to be a key part of the clean energy transition. While that conjures up images of pylons doing yoga in fields,* it actually refers to electricity storage, flexible demand, smart meters, and using interconnectors.[24]

Storing excess renewable energy is going to be absolutely key to helping solve climate change. Batteries will play a role, and if there's anyone that works tirelessly, it's the Duracell bunny. Tesla has also been setting up Gigafactories, which I assume are staffed by small orange men† with Elon Musk as essentially a Willy Wonka-type character. In 2017, Tesla produced the world's largest battery in Southern Australia. The Hornsdale Power Reserve – or, to give it its full name, the AAA $\times$ 10^{-12} battery – is attached to a wind farm and can power 30,000 homes for twenty-four hours a day.[25] Batteries are getting bigger and bigger all the time. Already a new one is being developed in Kurri Kurri, Australia, that's eight times the size of the Tesla one.[26]

---

* Dibs on starting this Instagram account.

† Australians.

There are other novel storage options too. A friend of mine works for a Scottish company called Gravitricity. I asked him about what the company is doing and, although I got distracted by how many times he said the words 'shaft' and 'deliver a load' in the conversation, the gist was that they are essentially dropping a heavy weight down old mine shafts to use old apple-man Isaac Newton's law of gravity to generate power. It's similar to the way pumped hydro works by pumping water up a mountain, and releasing it at peak times to meet demand. Again, old fossil infrastructure being put to a clean use.

Your-anium

Pretty much everything I learned about nuclear power I learned from *The Simpsons*. I cannot see a nuclear station without assuming it is run by Mr Burns and his assistant Smithers, and operated by employees like Homer Simpson. My only other reference is the TV show *Chernobyl*, which felt as harrowing as watching the latest season of *The Simpsons*.*

Lots of people are against nuclear power because they think it's dangerous, or link it to nuclear weapons. I believe this has done much to skew people's understanding of the technology. And sure, there have been incidents, like Chernobyl in 1986 and the 1979 accident on Three Mile Island – the most literally named place in the world. Obviously, Fukushima in Japan in 2011 was also very serious. Interestingly, the nearby

* I accidently added the *Chernobyl* soundtrack to our birthing playlist, but thankfully noticed in time.

Onagawa nuclear plant faced similar conditions to Fukushima but remained largely intact due to impressive safety culture around tsunamis, including constructing reactors higher than they needed to be.[27]

But nuclear power is also a well-established option for providing stable, low-carbon electricity – assuming you can get your hands on a nice supply of uranium. France is about 80 per cent powered by nuclear energy, and has been since the 1970s. (It's surprising France has got rid of their fossil fuel power stations when they still think smoking is sexy.) However, nuclear takes a long time to build and its costs are proving to be quite high, especially in developed countries. One main issue is that, unlike, say, solar and wind, you don't build loads of separate nuclear plants, and so the economies of scale don't really work to bring costs down.

The tide is high

There are many other types of renewables that haven't taken off yet due to technological issues, prohibitively high costs, or physical constraints, such as location. Wave and tidal power are technology options that have never been rolled out on a large scale but could have benefits over wind and solar because they are much more predictable and consistent. If we can ever get these scaled up, they could provide stable base-load power. Also, when you install them, you get to go surfing. Islands around the north of Scotland, like Orkney, have some of the best tidal potential in the world. It's just about making the economics work.[28]

Finally, there is simply using waste, and anything else we can find to create power. In Seville, they have so many oranges falling off trees around the city that they are piloting a scheme to use the methane the oranges generate when fermenting to make electricity.[29] The same could be done with other fruits. When life gives you lemons, use those lemons to power a hospital.

Human battery farms

What can we do about electricity at home? Sticking solar panels on your roof is great if your roof gets lots of sunlight and you have the money. It's not like those slates are doing much else beyond providing a place for birds to poo. Having just bought my first house, I will eventually get around to installing solar panels when I have the money, or during the next pandemic/ retirement. For the time being, we have some little garden solar lights, which are great if you want to stand and be lit up in one specific part of your garden at night like a maniac. If you rent or can't afford panels, changing your energy supplier to a renewables tariff is a good option.

Scientists have even found a way to use the temperature of human bodies to create electricity – yes, just like in *The Matrix*.[30] The hope is that our watches won't need batteries, as we will power them ourselves. That way you can sit around reading this book on your Kindle while powering it yourself. I asked my electrician mate Ian if he thought this was the future of electricity – human battery farms? No. Obviously not, he replied. But he did say that being a living battery, having your energy drained from your limp, lifeless body, sounded a lot better than, say, your wife leaving you for a twenty-four-year-old CrossFit instructor. Food for thought, isn't it?

15
Birth

I choose to birth and parent with love

People always say that the birth of their child was the happiest day of their life. I'd say the happiest day of *my* life was a toss-up between my wedding day and any time I've been to a waterpark. The birth of my son was perhaps the single most traumatic day I have ever experienced.

It started in the middle of the night, although my wife kindly didn't wake me up straight away. She gently prodded me awake at 5am to say, 'I've been having contractions for the last three hours.'

Never have I felt more instantly awake.

We went through all the steps: tried to eat and prepare ourselves, texted our family, etc. It seemed hard to monitor the contractions as they appeared to be quite sporadic. We stuck on *Four Weddings and a Funeral* to try to distract ourselves. I think we managed to watch twenty minutes over the course of several hours. We had been watching a lot of romcoms over the previous week, as they were all we could handle. After a few calls back and forth to the hospital we finally decided to make the thirty-minute journey in.

This was the bit I had been dreading, but thankfully the car started fine, and I remembered how to get there and how to drive. I cursed at traffic lights and the fact I took a wrong turn down a country lane while my wife listened to breathing techniques on her headphones, though she was nauseated and in pain. Importantly, we arrived without her having to give birth in the back seat of a Renault Clio.* I helped my wife to the door of the hospital, as I was not allowed to enter due to the pandemic restrictions.

Then things started to go wrong. She wasn't far enough along to be admitted, but they also had some concerns about her blood pressure, so she wasn't allowed to leave. I thought I was saying goodbye to her for ten minutes or so. Now we were stuck in this seemingly endless limbo where she couldn't leave and I couldn't come in. I sat outside in the car park for six hours, unable to enter the hospital while my wife went through a difficult early labour alone. I am not sure how many times I cried during this time. I tried to distract myself by reading a book about climate change. That was a fucking stupid idea. I went for a pee behind the car at least three times. I spoke to her on and off. I looked at pictures of us on my phone. I scrolled through Twitter, and people were going about their day, talking about how everything was a cake. Many months later, my wife pointed out that, at that time during the pandemic, the pubs were open. Technically, I could have gone for a pint to celebrate the birth, but I wasn't allowed to hold her hand through it – clearly the government's priorities lay elsewhere.

* I will explain the whole new car situation in the next chapter.

Finally, I was called in. We were both already exhausted. We hoped it would be over soon. Turned out we were only halfway there. I won't go into much more detail than to say we never really got the birth we wanted. During the next hours, I tried to remember as many of the hypnobirthing techniques as possible to help my wife with her breathing and to help her feel relaxed. There was a point in the very early hours where she and I were both falling asleep between contractions that were, at best, a minute apart.

That's not to say that the outcome wasn't wonderful – it was – but when he arrived at almost 8am the next day, what I felt most acutely was relief. Relief that it was over for her, and that everyone appeared to be safe. I had the honour of cutting the cord under significant duress from some nurses.

'Do you want to cut the cord?'

'No, thanks.'

'Why doesn't Dad cut the cord? Go on.'

They made it out to be more like a ribbon-cutting ceremony at the opening of a new store than a weird surgical procedure that they were suddenly thrusting on a man who hadn't slept for two days.

The next hours were beautiful. Within moments, I could see my wife was an incredible mother. We all bonded. The euphoria lasted about seven hours before I finally succumbed to sleep for an hour on a beanbag in the corner of a hospital room.

I remember holding him in my arms when I got back home and feeling in awe of him. That he was now part of our wee family, and that this tiny person was going to be with us for the rest of our lives. I thought back to the time in the weird Airbnb nine months ago, when we'd found out about the pregnancy. I had been underwhelmed then. Now, I was *over*whelmed.

When I was born in 1985, climate change was nascent in the public consciousness. *Back to the Future* was on cinema screens and atmospheric carbon dioxide concentration was a mere 346 parts per million. That concentration is pretty much what the climate organisation 350.org believe is a sensible target for humanity to consider aiming for – hence their name. As cinemagoers watched a young Michael J. Fox using a car to go back in time thirty years, little did they know what future lay ahead for them over the thirty years to come. Although scientists had already understood the issues for decades – and that included internal scientists at many big fossil fuel companies – the climate issue only properly came to the fore globally for the first time in 1988. My little brother had just been born and I was three years old. I devoted the majority of my energy and time to Thomas the Tank Engine; an obsession with trains that has since returned with my climate-conscious ways. Back then, there was considerably less of a partisan divide on the issue. That year, George W. Bush Senior famously said that 'those who think we are powerless to do anything about the greenhouse effect forget about the White House effect'. Hardly a tree-hugger.

The next year, the UN Intergovernmental Panel on Climate Change was formed to bring together all the science so the world could speak with one voice on the issue. Their first report came out in 1990. In that same year, the UK Prime Minister Margaret Thatcher noted in a speech that: 'The danger of global warming is as yet unseen, but real enough for us to make changes and sacrifices, so that we do not live at the expense of future generations.' All I really remember about Thatcher is that she was the person my grandmother would boo every time she appeared on TV, and I'd join in with a chorus of 'Maggie, Maggie, Maggie, Out, Out, Out!' I didn't know what it meant.

I just thought she was another old lady that my gran didn't like, just like Jean Garscadden across the road in number ninety-two. I knew Gran didn't like Jean because Jean never trimmed the hedge out front, and it made the street look 'unsightly'. I had no idea at the time why she didn't like Maggie, though. Still, at that time, it was forward thinking. Perhaps, this soon after Chernobyl, it was easy to imagine the idea of coming together to combat an existential threat to the world.

The nineties, therefore, began with considerable (and, it turned out, naive) optimism that the issue would soon be done and dusted. Some of this probably came from how swiftly the world had taken action to protect the ozone layer following the Montreal Protocol, where Ozone depleting gases were swiftly banned. The same would not happen with climate change. Negotiations began at the Rio Summit in 1992 and led to the Kyoto Protocol in 1997, the same year I went to high school. Many believed that this would quickly solve the issue. As we now know, this could not have been further from the truth. It was the beginning of the beginning. The next decade or so saw a combination of factors that led to inaction and pushback on doing anything to stop progress on the issue. 1998 was one of the warmest years on record (and also the year that I first kissed a girl).

In 2001, as I was sitting my exams, *The Fast and the Furious* came out, and it was the second-warmest year ever on record. It is now only the twentieth-warmest year (and probably only the twentieth *The Fast and the Furious* film). In 2006, when *The Fast and the Furious: Tokyo Drift* skidded on to our screens, China emitted more than the USA for the first time – and another film, called *An Inconvenient Truth*, became the way many people found out about the climate problem. A sense of urgency was born. That year, I also graduated from university (for the first time) and got a job in finance. By the time I'd started my PhD in 2009, the urgency was almost extinguished, as we had both the Copenhagen Climate Summit and a film simply titled *Fast & Furious*, both of which were massive disappointments. In 2013, I graduated with a doctorate, saw a sixth *The Fast and the Furious* film, and the level of global annual CO_2 emissions were now 75 per cent higher than they'd been when I was born. In 2015, it was *Furious 7* and the Paris Agreement – both excellent. And by the time *Fast & Furious Presents: Hobbs & Shaw* found a place in my heart in 2019, I was married.

I guess what I'm saying is that, during the short time that I have grown from a baby to an adult and had a baby of my own, and experienced everything I have ever experienced in my life, including all of *The Fast and the Furious* franchise, over 55 per cent of all the CO_2 emissions that have ever occurred, have occurred. I cannot let this happen to him.

Oh, and we named him. I will keep his name to myself. He rarely gets called by his name anyway, and is more often known by the array of nicknames he quickly gathered in our house. From here on out, I will simply refer to him by the name I suggested that was instantly vetoed by my wife. The year Oscar Winning was born, there was a fifth more CO_2 in the air we breathe than there was when I was born in 1985.

The question still remained in my head: what sort of impact would climate change have on the rest of his life?

16

Are cars the worst?

'Round, round, get around, I get around,' sang the Beach Boys. This actually wasn't about going down the Queen's Head and getting in two pints of Stella and a vodka lime, but about human mobility. We all get around an awful lot nowadays, while the world has figuratively got smaller. All with a serious impact on the planet. Personal transport tends to count for around a third of the average person's climate impact in countries like the UK and USA, and motor vehicles are the mode that dominate how we get from A to B in our daily lives.[1]

I've just bought my very first car aged thirty-five. (Me, not the car.)

It must be said that I am not a particularly big fan of automobiles. I get that people like them, but to me they're just a thing that you use for practical reasons, like the stairs. I don't really understand why, if there are magazines about cars, *What Stairs?* magazine doesn't exist. I hated taking driving lessons more than anything in the world, and felt a physical pang of anguish when I heard the horn honk outside to signal my driving instructor had arrived. This has stuck with me like a Pavlovian response, and every time I hear a horn, my body tenses in a wave of distress. Which is kind of the point of horns, anyway, so I guess that's fine.

Despite all the horrible lessons, I failed my test for going too slowly. I wasn't sure if a short stretch of road was a sixty-miles-per-hour zone or not, so I stayed at thirty. I failed by being safe. I am genuinely still not over this. Out of sheer stubbornness, I refused to take the test again for seven years. Later, I shared a family car with my brother for several years, which we both neglected due to a lack of interest, and I avoided because of fear I would lean on the car horn by mistake and trigger off a panic attack. I just remembered that my dad made my brother take his driving test in a suit. Yes, exactly. He thought that it'd make him seem like a respectable seventeen-year-old, and not like all the other boy racers trying to pass their test with no respect for the Highway Code. He failed.

Eventually I moved to London, where you don't need a car, despite it being full of cars. I sometimes wonder where all the people in cars are driving to in London, and I have a feeling if you stopped them and asked, they'd say: 'I'm not sure really – it's just what we've always done. Now, brmmm brmmm away!'

We moved out of the city two months before my son was born. There was no quick public transport access to the hospital, which was thirty minutes away by car, and we were both dead against the idea of walking there while my wife was in labour. I was consumed with worry about transport and becoming a new parent. Would we be able to get to the hospital quickly enough for the birth? What if he needs to be rushed back to hospital once he's home? I even worried about getting him to sleep without a car, because my dad used to drive me around to get me to fall asleep as a baby (so my emissions were sky-high before I was a one-year-old).

So, I finally relented, after all this time, and did some research into becoming a proper 'motorist'. Then I turned off the film *Cars*, because it wasn't as helpful as I had assumed. *Cars* 2 was even less helpful – you'd think I'd have learned my lesson by then. All I really learned from all that research was that *Cars 3* is really underrated.

I decided instead to ask a transport expert. Jillian Anable is a professor at Leeds University who, in her own words, studies: 'How to improve our messed-up transport system. You know, the one where incredibly valuable boxes of highly engineered steel and plastic sit empty and unused for all but one hour a day, and for that hour are often stationary in a traffic jam with one person sitting in them; or where it costs more to travel five minutes down the road on a bus than buy a triple-shot grande latte.' It turns out people who study the backwardness of our transport system need to have a sense of humour to cope. She pointed me in the direction of some useful research.

Cars everywhere!

There are about 40 million vehicles in the UK, and around 67 million people.[2,3] That is almost two thirds of a vehicle for every single person – like one of the cars from *The Flintstones*. All these folk worried about immigrants taking over, and it's the cars they should have been watching all along. It is estimated there are over a billion cars in the world. That's so many cars, that if you lined them all up, end to end, you'd either be asked to stop doing that or given the Turner Prize. I only started noticing how many cars there are quite recently. Most of them are doing nothing. The average car is only used 4 per cent of the time. A third of private cars are not used every day, and 8 per cent of them are not used at all during the week.[4] Lazy buggers. And while it's better that they're doing nothing than being driven, it'd be even better if they didn't exist. Cars have come to dominate the physical landscape of the societies we live in. Everywhere outside, there are vans and buses and red cars and blue cars and parked cars.* And inside, it's adverts constantly on TV selling you new cars, soundtracked by *The Cars*, in between shows hosted by Jimmy Carr and Alan Carr.

Road transport is damaging on two environmental fronts. Globally, it is a significant cause of greenhouse gas emissions (GHG), and locally, it is often an enormous source of both harmful air pollution and people getting angry and screaming 'Are you blind, open your eyes, Mr Magoo!' at each other. The overall GHG emissions come partially from the manufacturing stage, but mostly from the fact there is a combustion engine in the front of the vehicle, which is essentially a set of mini explosions from burning fossil fuel, and that's what propels it forwards. Emissions from road transport are rising in developing countries, and are often the largest source of emissions in richer countries. In the UK, emissions from the electricity and industry sectors have reduced significantly since 1990, but transport has remained stubbornly there, a bit like Jeremy Clarkson, hardly changing at all and still causing significant

---

* This is genuinely how I describe cars to people in conversation. No, I don't know what your Hyundai looks like, Ian.

harm. Transport is now the largest contributor to UK emissions, as improvements in efficiency have been offset by more cars and more driving.

The diesel and SUV debacles

It has become confusing trying to understand which type of combustion engine car is the worst – petrol, diesel or simply any with a personal registration plate. Diesel cars were pushed as more climate-friendly than petrol cars for a time, as they tended to produce less GHG emissions, and they were given tax breaks which made them pretty popular. However, the trade-off was greater levels of nitrogen oxide (NO*x*) and fine particulate matter ($PM_{2.5}$), which cause local air pollution.* Then, in 2015, the so-called 'Dieselgate' scandal occurred in the auto industry. Essentially, Volkswagen were fixing their laboratory tests to show low NO*x* emissions so their diesel cars would pass required standards, while in real life driving them on the roads caused much higher emissions. Around 11 million cars had these cheat devices installed, including 8.5 million in Europe.[5] Customers therefore thought they had much cleaner cars than they actually did.

The scandal has cost Volkswagen over 30 billion euros already.[6] It was a real blot on the history of Volkswagen – which,

* Air pollution is a different, though closely related, issue to climate change. Air pollution tends to be more localised and include lots of different pollutants that cause environmental harm to humans. An example is soot which is bad for human health. They both interact with each other and often things that cause climate change also cause air pollution and vice versa.

considering it was founded by the Nazis, is really saying something. Interestingly, the city of Cambridge, Massachusetts, has become the first US city to mandate labels on petrol pumps similar to those on cigarette packets, with yellow stickers warning of 'major consequences on human health and the environment, including contributing to climate change'.[7] I think it should go all the way and, as with cigarette warning labels, add images on the side of the pumps showing the damage caused by climate change. Although fitting almost 8 billion people in one image may be quite tricky.

The trend in the last decade or so towards greater ownership of SUVs has made things much worse for the climate. I believe SUV stands for 'Such Unnecessary Vehicles', which should be said like a contestant from *Ru Paul's Drag Race*. We have gone from 200,000 SUVs in the UK in 2010 to almost a million a decade later.[8] On average, SUVs require a quarter more energy than a medium-sized car,[9] and they have provided the second-largest contribution to increased global emissions since 2010 (after electricity). This has been the culmination of a trend where cars, like my waist size during the COVID lockdown, have become unnecessarily bigger and bigger over time. A third of new cars in Europe and half in America are SUVs.[10] SUVs outsold electric vehicles in the UK by thirty-seven times in 2018.[11]

Professor Anable says that, 'There is a clear trade-off to be made: the more we can shrink the size and weight of the cars we drive, the less we will have to restrict how much they are driven.'

These big bastard cars have become a daft status symbol in our society. You'd never find me wasting time and money on acquiring a status symbol, or my name's not Dr Matt Winning,

PhD. This trend for bigger cars seems to be another unnecessary American import. SUVs are also considered good for 'safety', but that very much depends upon whose safety you are talking about. A bit like gun ownership. "We need to protect our kids". Regardless of the fact that you can barely see the road and are driving a six-year-old to school in a small tank.

Many seem oblivious to how an idling SUV outside the local primary school is affecting the kids' health and contributing to a climate-altered future for them. However, understanding of the impacts is beginning to catch on with increasingly concerned parents. Groups, such as Mums For Lungs in London, campaign and raise awareness about the toxic pollution on city roads and the impact it has on child development. There was also the time when a bunch of badass schoolchildren in Tameside near Manchester dressed up as cops and handed out fake parking tickets to idling parents in cars around the school, after they found out that air pollution levels breached legal limits.[12] Which is fantastic and made a huge difference – just as long as they don't grow up wanting to be traffic wardens.

While it's easy to blame consumers, the problem is that these humongous cars are being pushed on them like supersized meal upgrades. SUVs are the most profitable cars to sell for manufacturers, and therefore attractive financing packages are made available, and spending on advertising has increased. In France, carmakers spend 40 per cent more of their budget marketing SUVs than normal cars.[13] A recent report by the New Weather Institute think tank and climate charity Possible, has called for a ban on advertising large cars due to their high pollution. We don't need such big cars. A 2020 report found that 150,000 new cars in the UK were too big to fit into standard parking spaces.[14]

Personally, I measure whether or not a car is too big by whether

I could fit it into a tent. Bear with me. The year was 2012 and I was performing comedy at a music festival near Loch Ness, which was wonderfully named Rockness. Lots of comedians were hanging out in a designated backstage camping area, and I was in a mischievous mood. One comedian, Ray Bradshaw, and his mate had an enormous tent, and so I hatched a plan to see if we could fit someone's car inside it while they were gone. I cleared their belongings from the tent and then helped the driver reverse the car into it. It fitted perfectly. We then zipped up the tent and left it there. Word got around, and when they returned to the artists campsite to fetch a jumper later that evening, there was an enormous crowd that 'just happened' to be there. Needless to say, they got quite the shock. It remains the single funniest thing I've ever done or said to date. Much better than this book.

What about if you already have a car?

There is still a bunch of stuff you can do to limit your emissions from driving. Firstly, drive at the right speed. While 88 miles per hour might be the ideal speed for travelling back in time, it is not a good speed for anything else. Most people on motorways want to get where they are going as quickly as they possibly can. But this comes at both a financial and environmental cost, because it consumes more fuel to go the same distance as the car has to work harder against wind resistance at higher speeds. Reducing the upper speed limit to fifty-five or sixty miles per hour would actually be pretty effective in cutting emissions and will save you money.

Not idling in a car is another piece of advice. If you're stationary for more than about ten seconds, it is better to switch the engine off. Drive-throughs are, therefore, an inefficient practice. Just go into McDonald's, people. I know it feels magical talking into a box and then having your food delivered to your car out of a window. But if you really need your fix, then do the classic drunk-walk-through-the-drive-through-pretending-you're-a-car.

Keeping the car light is good too. I don't mean *bright* – that'll kill the battery. I mean not full of dumbbells. Obviously, if its full of people, that is better as you'll be sharing the emissions, but extra unnecessary weight reduces efficiency. And keep your tyres inflated, as anything under the recommended PSI also reduces efficiency.*

Sharing journeys with others is an obvious solution. Most cars on the morning commute contain one person – usually the driver – and four empty seats. This is an enormous waste of resources and money. Now, I understand that everyone likes their own space, especially when they are still waking up, and doesn't necessarily want a car full of randoms to make awkward chit-chat with. But if you could ride-share a day or two a week, it will cut your impact on the climate. Better to share a car now than a toilet in an underground bunker in thirty years' time, eh? Websites such as Liftshare and BlaBlacar provide opportunities to arrange this. Another ride-sharing solution could be turning driving lessons into a commute: you may even luck out and get a lift to work from a seventeen-year-old boy wearing a suit.

Car clubs have been growing with apps such as Zipcar and Co-wheels. These appear to be especially popular amongst younger generations. The idea of 'ownership' seems to be an old person's aspiration, whereas the younger generation just

* WHY are the screw caps on tyres so fiddly and easy to lose?

want the 'service' of getting from A to B. In our last year in London, I had a Zipcar membership, which helped with late-night trips to A&E. (I'd love to say A&E was a trendy club in Shoreditch, but it wasn't. On the plus side, it wasn't full of twats and the drugs were much better quality.) A US study suggested that people who start using such car-sharing schemes reduce their transport emissions by 51 per cent.[15] The problem is that they are only available to a relatively small part of the population, mostly in bigger cities, and these schemes have not put a dent in car ownership at all. The end goal for all this may be the development of peer-to-peer sharing, where we basically all put our cars on an app so they can be rented out, just like Airbnb.* If you had access to all the cars outside on your street, you'd probably start thinking twice about whether you needed one yourself. Plus, you could find out what your neighbours listen to on the radio.

Overall, though, it won't be enough. Professor Anable says the most important thing is to encourage people to use their cars less. But that is difficult, she says, because people are addicted to their cars, and when she talks to them about using their cars less, she says they look at her like she's 'suggested they should drink their own piss'. She proposes setting up the first Car-aholics Anonymous groups in the world. 'Start people on the seven-step programme. Once they've mastered this, there's the twelve-step programme, followed by the twenty-five-step programme. And once they've done twenty-five steps, the car-aholics may even find themselves able to walk as far as the local shops.' So, like meat-free Mondays, maybe we need clean-transport Tuesdays?

---

* Air A2B?

Get on yer bike

The flip side of this is that making better public transport that people want to take is vital. Public transport will always be on the back foot against having our own modes of transport, because fundamentally, deep down, we dislike sharing with other people. To gain traction, it needs to be significantly cheaper and more convenient. That's a tough ask. Though, if you look hard enough, there are all sorts of great deals to be found with bus travel. For example, Megabus now, as standard, throw in, at no extra cost, post-traumatic stress disorder. Local trains need to be much cheaper. Trams in cities are similarly helpful, although they are essentially buses that can only go in a straight line.

Active travel is the fancy name given to walking, cycling, e-bikes, etc. in an attempt to make climate action appeal to people who work for fitness brands. These options have the added benefit of being good for your health – and your wallet. (Those who partake in active travel often have to put up with the second-hand pollution from cars, although this will decrease somewhat with electric vehicles.) Cycling short journeys in cities is an obvious quick fix. A recent study by the UK Energy Research Centre suggested that switching just one trip a day from driving to cycling in cities can reduce a person's emissions by half a tonne of CO_2 a year.[16] My friend Ian and his wife once bought a tandem as a sort of quirky, romantic thing to do. They're divorced now. He cycles it to the shops by himself. It is one of the saddest sights known to man: a single person on a tandem. Maybe he needs a tandem ride-share scheme.

To EV or not to EV, that is the question

What about if you're thinking about buying a new car? The first question is, do you really need one? While the idea of a 'car-free' society is probably fantasy, we do need to reduce the number of cars overall. This is hard. If we can't talk about what combination of A-roads and motorways we took to get there, then we'd have literally nothing to say to other people at weddings.

If you are going to buy a car, then what should you get? Well, it turns out that Elon Musk was right about one thing. No, not about the scuba-paedo thing or COVID-19. He was right that the future is electric vehicles (EVs). The cars we drive are changing. The next *Transformers* film is just going to be three hours of Optimus Prime recharging himself. Estimates suggest that sometime around the mid-2020s, EVs will be the cheapest option on the market everywhere, even without government subsidies.[17] It is expected there will be forty million EVs in Europe by 2030,[18] and Ford plans for all its European sales to be electric by the same year.[19]

Some countries are already way ahead. In Norway, more than half of new car sales are electric[20] due to clever government tax incentives making them cheaper. Funnily enough, the Norwegian pop sensation A-ha were early adopters of EVs in the 1980s.[21] I assure you this is not a joke. Two of the band bought a Fiat Panda in Switzerland in 1989 that had been converted to run via battery, and shipped it back to Norway. It did about 45 kilometres per hour. As no rules existed for electric cars, they made them up themselves and registered it as a diesel motorhome. They then drove about in it and refused to pay any road tolls, as they thought there should be incentives for more people driving electric cars. They were given tickets for this, but refused to pay the £30 fine. When the car was then confiscated, they simply bought it back at auction for £20 since nobody else wanted it. Then the biggest pop stars in the country went straight back to driving it about Oslo, refusing to pay their road tolls once more. This happened on numerous occasions. Since then, EVs in Norway have been exempt from a hefty registration fee and road tolls. This forward thinking from the seminal eighties synth-pop group is credited as playing an instrumental role in Norway's early EV uptake (on me).

The UK has committed to banning the sale of new petrol or diesel cars from 2030 onwards, which is a great step in the right direction. The only issue I can see is that if you ban Diesel from cars now, then the next *The Fast and the Furious* film will be terrible.

One classic argument that you hear is that 'there are emissions from electric vehicles too!' Some people suggest that maybe they're worse because of the batteries, and because the electricity might come from fossil fuels. This argument is usually made by someone's uncle at a family party, who has done zero research, or by traditional car companies who pay for some rather shoddy 'research'.[22] The simple fact is that, at present, emissions from an electric car in the EU over its lifetime are on average about three times lower than the equivalent petrol or diesel car, over their lifetime.[23] The production stage of an EV does tend to be worse emissions-wise than an internal combustion engine (ICE), but this is more than compensated for by the emissions saved during driving. And EV emissions are only going to reduce as the electricity system decarbonises.

Unlike with Bob Dylan, the public are going to love their cars 'going electric'. James Beard, a seasoned and knowledgeable climate campaigner, told me that the switch to EVs will provide a much better experience for consumers, because they won't need to be serviced as much.[24] James says: 'Turns out, if you don't have to operate a series of controlled explosions in your car on a regular basis, it's easier to maintain.' Personally, as someone who has zero knowledge whatsoever about the ins and outs of cars, it does appeal somewhat that I won't constantly feel out of my depth talking to a mechanic about fuel pumps and carburetors.

He stresses, though, that this means a significant workforce won't have much to do any more, which is a problem. As for other jobs based around the use of fossil fuels, this requires far more thought about how to fairly transition workers to new jobs. In the UK, we will not be as badly affected by this as other countries, like Germany, that make the engine parts. Here, we mostly make stuff still required by EVs, like doors and windows.

Various new cars are being specifically designed from scratch to be electric rather than just trying to cram everything into the current ICE car design as happens at present. The new design by Tesla is basically starting from the bottom up, and Volkswagen, too, are taking this route now.

The main issue is, of course, charging. Basically, if you have a driveway, then this is not going to be a problem. Your life will simply be better, because you will never again go to a petrol station and have to act cool when going to pay for petrol at the kiosk, repeating 'Pump 8, pump 8, pump 8' to yourself over and over in your head, but thinking, 'Just say it casually when they ask,' and then, as you approach the counter and the person says 'Any petrol sir?', you just scream 'PUMP EIGHT!' in their face. In fact, you may be able to use the car's battery to power your house at times if needed, or you could rent out your charger to others who don't have one. Pretending to own a petrol station was a game I fondly remember my brother and I playing as wee boys. Now, as he owns a driveway, he can live that dream.

However, if you don't have a driveway, like me, there are going to be a whole host of other options, like lampposts, supermarkets and work car parks, etc. Tesco has committed to installing free chargers at 600 supermarkets across the UK.[25] It will essentially be like how cafés use free wi-fi to entice you in. Scaling up the charging network will be essential over the next few years, though the reality is it's already better than most people realise.[26]

There are also halfway-house options, like plug-in hybrid electric vehicles (PHEVs) and self-charging hybrids. Understandably, some people can be hesitant, not wanting to jump straight from one extreme to another, and preferring a transitional step in between. And, once upon a time, these hybrids were the greenest options in town, because nobody was making EVs. Now, it appears that hybrids are often the worst of both worlds. PHEVs can be incredibly green when running on a battery charged by 100 per cent renewables. Otherwise, they are incredibly inefficient.

'When they are not running off their battery, they're the most inefficient vehicle you could possibly design,' says Beard. The extra weight of the battery means they require more petrol, creating astronomical fuel costs and more pollution. If you get a full-battery electric, you'll probably need to charge it once a week, or once a fortnight, says Beard. 'If you get a plug-in hybrid, you need to charge every other day to make sure you're using the battery and not the engine.' A lot of PHEVs on the roads are company cars, too, so many of the people running them are not paying for the fuel and don't care about all the extra emissions.

Having a family appears to be an obstacle to overcoming our car obsession.

In the end, I bought a second-hand petrol car. I couldn't buy an electric car because I didn't have the money and also, I'd have nowhere to charge it as a cable would have to go several feet out of my house and across the pavement, if I was even lucky enough to get parked outside my own house.* The four days I spent researching cars were the worst days of my life. Now I own a Renault Clio. It has the smallest engine I could get and slows to a crawl driving up a hill. After my son's birth, the drive back from the hospital was easily the most terrifying journey of my life.† I'd had about an hour's sleep, and was navigating through country roads in the dark back to a house we'd only lived in for two months with the smallest person I'd ever met in the back seat.

I wonder whether my son will ever have to learn to drive. In the year 2037, will there only be driverless cars, like that terrifying kids' TV show *Brum*? If KITT from *Knight Rider* is essentially going to be the norm, then we'd all better start growing our Hasselhoff chest hair now. I hope my son won't be as scared of driving lessons as I was, if only because they

---

* Though I did consider setting up a Hawaiian bar in my house and making pedestrians limbo under it.

† And I've taken a taxi in Greece.

won't exist, and the practical driving test will simply be the bit where you read a registration plate from a distance. Then you'll download an app and that's it. Passed. Though the DVLA will still charge £100 to sit it.

17

Can I still go on holiday?

It goes without saying that going on holiday is wonderful and pretty much the one thing most of us live for, because we get to escape our mundane lives, awful jobs and mad country for a while. If, like me, you live in the UK, then getting off this soggy, grey island for a week or two to stare at a giant red ball in the sky through some £4 H&M sunglasses with a cold *cerveza* in one hand and a load of coins that you don't understand in the other is the absolute pinnacle of your year. Not even the regret of all-over sunburn on the first day can get you down.

However, flying is probably the most carbon-intensive activity you can undertake – and to 'sun, sand and sex', you can also add CO_2 emissions. A return flight from London to New York causes about the same emissions as the average person from Paraguay creates over the course of an entire year.[1] And a return flight from London to Berlin is the equivalent emissions of not recycling for three years.[2] So, do we all need to give up our precious annual holiday to the Algarve, or is there perhaps something a bit easier for people to get on board with (no pun intended), like creating a race of superhumans who can fly?

Going on holiday itself does have an impact in terms of pina coladas, hotel elevators, overpriced tat for your mother, etc. A 2018 study found that, overall, tourism is responsible for 8 per cent of global emissions.[3] But of this, by far and away the biggest impact is likely to be how you get there, particularly if you fly. I'm afraid it doesn't matter how many times you reuse your hotel towels, that gesture is a bit like pissing into a wind turbine – you're the only one that will notice the impact (and smell of piss). In fact, if you take more than one return trip a year by air, then flying will probably be the single biggest element of your yearly personal contribution to climate change.

Low-carbon airplanes! Shirley, you can't be serious?

Unfortunately, we do not have many alternative methods to planes for long-distance travel. Most of us don't have the time to paddle a canoe to Barbados and back on our annual leave – never mind the upper body strength, the ability to paddle a canoe, or the canoe. Long distance electric planes are some way off, and teleportation isn't yet a thing. (And even if it was, we'd just end up with lots of low-budget teleporters who'd lose your body parts: 'I'm sorry, sir, we've accidentally sent your testicles to Budapest, would you like some complimentary ones?')

Electric planes may well be possible for domestic travel or short distances in a decade or two, which is good news for travelling around Europe, although electric high-speed trains could do a similar job. But we need to reduce emissions *today*, and it is unlikely flights run on battery power will ever be able to take us across big oceans. So while some long-distance zero-carbon flights may be possible, it will probably require you to hop on and off interchanges at six different airports to take an electric flight all the way to Bangkok. So, back to the *Titanic* it is.

Using biofuels (growing plants to burn as fuel) is another often touted option and does have benefits. However, scaling this up is unlikely to happen quickly enough to help in any meaningful way, and it comes with many issues. In many places, growing crops to burn for fuel means competing for land used for crops to feed humans and animals. You may have to trade off whether you'd rather go on holiday or eat for two weeks. Biofuels have also been linked to mass deforestation, as trees are cut down to make way for growing crops – not exactly a climate-friendly situation.[4]

Shouldn't airlines be doing something about this?

Well, there have been many positive stories from airline and oil companies about their low-carbon credentials and new types of solutions.

Ryanair advertised that it is the lowest-emissions airline in Europe.[5] This is technically true, although at the same time it is also Europe's highest emitting airline. Confusing? Yes. The reason it claims to have the lowest emissions is mostly due to Ryanair cramming on as many passengers as possible, meaning that they have the lowest emissions *per passenger*.* Like claiming

* The old 'increasing the denominator' routine.

to be the least-bad mass murderer because, even though you killed the most people, you did them all at once.

Now, don't get me wrong, packing in as many passengers as possible is a good thing, and maybe you can take solace in that. When your legs are squashed against the seat in front of you and you've got a sweaty elbow in your side, at least you're 'doing your bit'. But it's also unlikely that people would take as many flights as they do if they weren't made so cheap by cramming everyone on like sardines. Unsurprisingly, the Advertising Standards Authority took issue with Ryanair's claims. In fact, Ryanair is now one of the top ten carbon-emitting companies within Europe (the first time any company that is not in coal has entered the top ten). You can think of this list as a sort of *Top of the Pops* for the apocalypse.

In 2019, a BP adverting campaign posed the question, 'Can a banana skin fuel your flight?' This idea is at best hopeful and at worst disingenuous greenwashing. While it is true that food waste can provide some of the inputs for biofuels, scaling this up is still a long way off and there won't be enough of it available to provide all our aviation fuel. Besides, I always thought it'd be the baked beans I *did* eat that might make me airborne rather than the ones I *didn't*.

The truth is that reducing the number of flights people take is the only viable solution at the moment. Yet the number of people flying is expected to double by 2037.[6] Unfortunately, the improvements in efficiency which reduce emissions are always considerably outstripped by the increasing number of flights being taken.[7]

Now, I have an obvious solution to reduce the number of people taking flights. Hear me out. Some people are afraid of flying, right? I say we just need to make *more* people afraid of flying. Oxygen masks down the entire flight. The only inflight entertainment available is the film *Alive*. Actual snakes on a plane. Babies crying the entire time. How are we getting them to cry? We're blowing smoke in their faces.*

I suppose, in a way, EasyJet and Ryanair are actually the most environmentally conscious airlines out there, as they've been trying to put people off flying for years.

Grounded prices and sky-high celebs

One major problem with flying is that it is far too cheap, even without taking into consideration the harm it does to the climate. Have you ever found yourself wondering how it can cost the same to fly to Berlin as it does to get the train to Manchester? And then just assuming, 'I guess that's the way the world is,' and going back to incorrectly inputting your password into a booking app? Actually, you're 100 per cent right – it really should not be that cheap.

Unlike the tax you pay for your car fuel, the aviation industry pays no jet fuel tax.[8] In fact, the Yellow vest protesters in France spoke about how the airline industry is unfairly getting huge tax benefits compared to motorists.[9] In addition, you don't have to pay VAT on airline tickets, yet you pay it on tickets to the cinema. These are concessions that were necessary about fifty years ago to help get the industry up and running, but the aviation lobby have worked hard to keep them. It's unreasonably cheap. I've really got to stop saying this, as I'm trying to dissuade people from flying, and instead I find myself going on

---

* Please make sure you've read the chapter on heatwaves.

about the LOW LOW PRICES. Somehow though, we need to make flying less attractive than spending two weeks in Bognor Regis. And without incentives to achieve this, we might need to wait for climate change to turn Bognor Regis into the Amalfi coast. It's a Catch-22 (which is a book about planes).

One proposed option is a 'frequent flyer levy' – an increasing tax on every flight you take in a year after your first.* For instance, you would pay some tax on your second flight, double the tax on your third, then double it again on your fourth, and so on. Being allowed one tax-free flight would help protect the yearly family holiday most people take, while making additional air travel more expensive for those who use it most – the wealthy. The reasoning behind this is that most people don't actually fly that much. In the UK, about 70 per cent of all flights are taken by only 15 per cent of people.[10] A smallish number of wealthy types are doing most of the flying and causing most of the damage – who would have guessed? – and it only seems fair they should pay more to solve the problem. Transport researcher Dr Giulio Mattioli explains: 'If we're trying to reduce emissions in a fair and just way, perhaps we should tax someone's third holiday abroad in a year before we tax someone else's home heating, or journey to work?'

The rich love flying. A recent report from clean transport non-profit Transport & Environment showed that private jet emissions in Europe increased by 31 per cent between 2005 and 2019.[11] Almost a fifth of these come from the UK. Private jets are five to fourteen times worse per passenger than a com-

* Your first return flight, that is – you are allowed to come back. It wouldn't make you stay in Alicante, although I guess it wouldn't be all that bad.

mercial flight, and fifty times worse than a train.[12] The authors recommend that, from 2030 onwards, only electric aircraft powered with green hydrogen and electricity should be allowed for private jet flights under 1,000 km. Taxing these flights could help pay for research into improving low-carbon air travel so it happens sooner for the rest of us.

This sort of affluent lifestyle is seen as influential, so much so that we now have people called influencers, whose job, as far as I can tell, is to bring down society, one Instagram post at a time.

A study looking at the flying habits of celebrities including Jennifer Lopez, Bill Gates, Emma Watson, Mark Zuckerberg and Paris Hilton showed Gates won the competition for most emissions, totalling 356 hours of flights and 1,629 tonnes of CO_2 in 2017.[13] Mostly because he took lots of international flights. This would be seen by his online followers (46 million on Twitter, 20 million on Facebook and 3 million on Instagram). Next were Paris Hilton and Jennifer Lopez. Emma Watson had the least emissions, at only 15 tonnes of CO_2, presumably because most of her flights were by broomstick.

Brits fly abroad more than those from any other country.[14] This causes a tourism deficit in the UK, meaning more money is spent by British tourists going on holiday, than is spent by tourists coming into the UK. So, less flying abroad would benefit the UK economy – and the people who make those seaside picture boards that you poke your head through.

Fun modes of travel

In 2018, I decided to start actively taking trains more and flying less. According to the European Environment Agency, train travel is about twenty times better than an equivalent flight, and although this can range enormously, the train always comes out better.[15] For the emissions of one return flight from London to Madrid, you could make the same journey six times by train.[16] The trouble is that train travel can be expensive in comparison to flights. Hence we need specific policies to make it more attractive and easier. For example, France is banning short-haul internal flights where the same journey could be taken by train in under two-and-a-half hours.[17]

In 2019, I took the train to France on a ski holiday, and also travelled by train to Italy for work. I actually prefer the train to flying, anyway, because on flights you are treated like you've moved into an old folks' home – and not just because you might die at any moment. Firstly, the food is all served at the same time, on a little tray, little table, little meal, num num. Whereas on the train, you have the buffet car and you can stretch your legs whenever you want. Plus, you have plug sockets, so I like to take a slow-cooker with me, and in just five hours I've rustled up a Moroccan chickpea stew for everyone in Coach C. Pinch of turmeric, dash of cinnamon, sprinkle of paprika. I mean, let's be honest: the only thing exploding on that train is your taste buds. Flights also have those patronising safety announcements: 'This is how you put on a seatbelt . . . and *this* is a whistle.' Whereas on a train, they just go: 'There's an axe, there's a window. Work it out.' That's my kind of transport. For my thirtieth birthday we took the Caledonian Sleeper train from London to the Highlands, and enjoyed sitting in the old-school bar, sipping whisky and pretending to be characters in an average Connery *Bond* film.

TV shows like *Race Across the World* have helped show how exciting it can be to travel without flying. At the very least, we need Generation Z to do more interrailing around Europe and less backpacking to Thailand. The good news is that young Brits have stated they'd be willing to pay a quarter more to make a holiday eco-friendly.[18] So, young people, if you want to go travelling to find yourself, then why not try finding yourself in Norwich? Sites like The Man in Seat 61 can help you plan your rail travel,[19] and train ski holidays can be arranged by companies like Snow Carbon.

Duty-free hell

Opposing airport expansion at a time of climate emergency seems like a sensible move. Building extra infrastructure to accommodate increasing air travel demand, when it could be spent on developing domestic high-speed rail, is counter intuitive. Several airports are now claiming to be reducing their emissions or going net zero. For instance, Airports Council International Europe, who represent about 500 airports, has set a net-zero target for 2050 but it won't include emissions from aeroplanes. Their target only accounts for about 2 per cent of emissions associated with owning and operating an airport.[20] Which is a bit like saying that you have the most vegan abattoir because the cows eat grass.

Plus – and I am 100 per cent fine with saying this – airports themselves are the absolute worst cesspits imaginable on the face of this green earth. If Hell existed, I imagine it would be a never-ending conveyer belt of queuing to check in while stressing about infinitesimal grams of luggage, then passing through security to find out if you've accidentally become a drug smuggler since packing your suitcase, then undertaking a quest through infinite aisles of the most expensive consumer items imaginable, which somehow convince you they are a bargain because of the words 'duty-free'. Eventually, you find your gate, only to wait for eternity to board. Finally, you get to board the plane, walking down to reach the end of the little walkway to the back of the plane, pulling back the curtain, only to find yourself back at the entrance to the airport and forced to start again. That is Hell.

Can't I just offset my flights?

A common touted solution for guilty flyers is to 'offset' your flight. Carbon offsetting is where you pay for a reduction in emissions to compensate for emissions you have generated from some activity. You can 'offset' your flight by paying a company to plant a bunch of trees. Sounds too good to be true, right? It is. Now, it is better than nothing, sure. I've done it myself. But it's no way to run a society long-term and essentially is a nonsense. Plus, it is hard to prove that the trees are planted without flying back across the world to check.

There are many issues with offsetting as an approach. Firstly, you have to be wealthy to do it. Secondly, it is extremely difficult to prove that the avoided emissions are additional or that they wouldn't have otherwise occurred. And are they permanent?

Thirdly, we need to reduce actual emissions today, not in the future, which is what a lot of the offsetting options do, i.e. how long does it take a tree to grow and capture the emissions over? I'd say as an individual you probably shouldn't offset. I think that unless you have stupid amounts of money then it is not worthwhile. You'd probably be better giving the money to a climate NGO or group to help change the system.

Therefore, offsetting is a much less certain way to have a positive impact than simply not flying. It is more plaster than prevention.

Peeing in the pool

I realise giving up flying entirely is often not possible for those with family on the other side of the world. Although I guess for some it might provide a helpful excuse. 'Sorry, I can't come to Auntie Pam's for Christmas, Mum – the climate, you know, nightmare!' However, only a quarter of international flights from UK residents are to see friends and family.[21] For everyone else, those cheap, long-weekend city-break flights need to be thrown in the bin. What's even the point of going on holiday for three days? Stag-do's abroad are an awful idea, even without climate considerations. And for those who take longer holidays, why not try holidaying somewhere local every second year, or taking the train? Maybe we need some sort of incentive scheme not to fly, like non-air miles where you get money from train travel?

One thing that helps cut down flights considerably is a global pandemic. It has been interesting to note how the world adapted and changed when there was an immediate danger. The idea of no one flying before the pandemic was unimaginable to most. The pandemic has shown that when we realise we are in real danger, we will stop. Now, with climate change looming over us, I personally hope that this will cause people to realise we do have the ability the do things differently. Business travel is one aspect that I think will change dramatically.

I haven't flown for three years now. I imagine I may fly again at some point, once our son is conscripted into the screaming-baby-on-a-plane club. Obviously, the idea of taking him to see the wonderful cultures and sights of the world is a joyous one, and something I'd love to experience (plus going to a decent waterpark). But when we do venture far away, it will certainly be infrequently. Refusing to take domestic flights unless absolutely necessary is a good rule.

In fact, in 2019, the number of air passengers in Sweden dropped in the same year as the *flygskam* movement, which translates as 'flight shame', took off,* and similar trends occurred with domestic flights in Germany.[22] This is not to say we should shame others for flying, not at all, but shifting social norms by sharing stories of how *we* ourselves have chosen to fly less is helpful. Some will defend their flying habits by suggesting that it is not worth focusing on aviation because it's small potatoes, and only responsible for about 2.5 per cent of global CO_2 emissions. However, this is misguided.

Firstly, due to the large warming impact of water vapour (such as through contrails) and other non-CO_2 greenhouse gases from aeroplanes, aviation is responsible for at least 3.5 per cent of warming, and even up to 5 per cent.[23] But most importantly, it is an industry where emissions are rapidly growing, rather

* Of course, pun intended.

than shrinking, with no signs of this changing anytime soon, meaning it will take up a bigger and bigger proportion of global emissions as other sectors reduce their own climate impact. Plus, as stated above, aviation is the preserve of the wealthy. As Mattioli elegantly puts it: 'Saying that it's just 2.5 per cent so we shouldn't worry too much about it, is a bit like saying "I know peeing in the swimming pool is bad, but as most people except me and my mates don't do it, urine will not be more than 2.5 per cent of the pool – what's the big deal?"'

18

Early weeks

This will be my greatest achievement

How were the early weeks of parenthood going? Well, it was relentless. I had no idea what day of the week it was, and the time of day was utterly meaningless. All I knew was that the night was dark and full of terror. And long – oh, so long. But our hearts were fuller than they had ever been.

A good friend of ours visited in the garden for biscuits and coffee when our son was only two weeks old. It was a late, crisp summer morning with enough sun in the sky to sit outside for a while and chatter with the baby all wrapped up. It was lovely to speak with an actual human being. This was the first person Oscar had met beyond masked nurses and his maternal grandmother. During the visit, our friend told us that her aunt had described having a newborn as 'the best kind of living hell', and I have yet to hear a more apt description. It has stuck with me. The best kind of living hell.

Now that I was clear about what impact my son would have on climate change (REMINDER: it isn't really something worth focusing on), I was now ready to turn my attention to what sort of impact climate change will have on my son. Except. I had barely thought about climate change at all. I'd either been too happy, busy or tired. Life was too messy. We were literally making it through minute by minute, a day at a time. 'New dad' Matt Winning was in the building, and 'climate researcher' Matthew Winning was absolutely nowhere to be seen.* This must be what it's like for the majority of normal people going about their daily lives: far too busy, and rightfully blissfully unaware, to focus on existential threats to humanity. How you could be expected, as a new parent, to consciously make green decisions is beyond me. This is why we need to strive to make any decisions that are beneficial for the climate the easy, default thing to do. Easier said than done, though. I wondered if, so far, having a baby had affected our personal emissions? On the one hand, we'd driven more in order to go to appointments. On the other hand, we'd barely had time to shower.

I was really nervous about returning to work after Oscar's birth, despite the fact that due to COVID, my commute now consisted of opening the door to the front room. It was considerably faster than my usual hour squashed on the Tube, although I still managed to be late almost every day. It also meant I could see so much more of Oscar. For that I am grateful. I cannot believe that for most, paternity leave means taking only two weeks off – I took six. After two weeks you can still barely function, even as the parent who didn't give birth or isn't breastfeeding. We spent the first week having to take daily trips to various hospitals (three different ones) to

* I like to think of the climate researcher part of me as my alter ego: the more serious Matthew Winning.

have various tests. And then we had some feeding issues and Oscar had to go back into hospital for a night. We were lucky that my wife's mother could stay with us and help; we know other couples with babies who had no help whatsoever during this time, and the father still had to return to work after two weeks. It is insane.

After my mother-in-law left, we quickly had to become a well-oiled and regimented team doing whatever it took to survive. I had bought some reusable nappies, but found very quickly that the hypothetical good intentions of the climate researcher Matthew Winning gave way to the real-life Matt Winning saying 'Absolutely fuck that!' when faced with a leaky, poo-covered baby. A wonderful family friend gave me a parenting book called *Commando Dad*, written by an actual ex-Commando. At first, I wasn't sure about it. Now I 100 per cent agree that army exercises are probably best for preparing to have a baby. It sure feels like you are undertaking extreme conditional training – with sensory attacks and sleep deprivation. Waking up in the pitch dark to the sound of screaming. My God, the screams. Why is he screaming? When will it end? I think my wife and I developed into a fairly robust force, although she was clearly the most strong and calm. I was glad I was going through this with her. I was both in complete awe, and quite envious, of her. Not only was she able to calm him down with feeding, but she also has a lovely singing voice and had found a specific song that calmed him right down when he worked himself up into a tizz. Whereas I mostly resorted to saying the word 'please' again and again in the hope that he would somehow master the English language and understand me. I do have a big nose that he liked to chew on, so at least that's something.

And the tough bits were so tough. Nobody tells you about cutting their nails. You are expected to trim something that you can barely see on the end of a tiny human finger. Of course, I accidentally cut his finger once. I'd count this as the second-most traumatic day of my life, after the birth. I cried for twice as long as he did. Despite the fact that the doctor was lovely about it and said that it was extremely common, and that she had done it too, I still spent several weeks beating myself up about being a terrible father and wondering if my son would ever look at me the same way again.

However, there was incredible news in these early weeks, too: my son saw Scotland qualify for a major football tournament. This was something I had not seen for twenty-two years; it took him two months. I mean, he didn't really see it. He mostly slept on me. And then, when we won, I accidentally woke him up and we skipped around the room together in joy/surprise. I mean, technically, he was born in England, but his allegiance is already with the dark blue jersey – of this, I am sure. I declared him the luckiest mascot in the world and swore to watch all Scotland matches with him, forever. (Also, he will not find out that England have a football team for as long as I can help it.)

Anyway, I was back at work now. The bad news was, I asked people around the office, and they said climate change was definitely still happening. Turns out it doesn't just go away if you stop thinking about it. At least I had finally tested that theory after a decade of pondering it on a daily basis. My return to work was like cautiously dipping a toe into the world outside our little household. It was quite the change for us all. Thinking about anything beyond my new family felt refreshing, but also alien. And 'climate researcher' Matthew Winning was slowly returning back on the scene. It would be interesting to see how he battled for space in this brave new baby world.

19

Shall I put an extra jumper on?

Dealing with radiators. It is a rite of passage every dad must go through. I decided to go full steam ahead, and replace a radiator in my son's nursery before the winter arrived. It lay on the floor for two months, along with my self-esteem and the remnants of my masculinity. Thankfully, I was saved by our neighbour, Joel, who kindly took pity on me and showed me what to do in about three minutes. He left me a key and said, 'Use this to bleed the radiator.' I think this means I have to call *him* Dad now. He is also a new dad, and I know I shouldn't compare myself to others, but I worry he's already more of a dad than I am.

Our house

People forget our homes and buildings are part of the climate problem. A bit like how they forgot about Dr Dre, and Eminem didn't even bother to remind them. There are an estimated

28 million homes in the UK, all of which need heating and hot water,[1] and about 68 per cent of the country's buildings emissions come from these activities.[2] And as much as I relish taking up that other dad mantle of telling people to put a jumper on when it's cold, there are only so many jumpers you can reasonably ask someone to put on.

At the moment, about 85 per cent of UK homes are supplied from the gas grid network, which pumps natural gas – sorry, fossil gas – into our homes.[3] The other 15 per cent are presumably freezing or warmed by the smugness of living *off-grid*. Overall, 44 per cent of UK energy consumption is used for heating.[4]

Fossil gas is burned to heat radiators, and these hot, white metal bastards subsequently heat the air in our rooms. How effective the heating is depends upon how leaky the building is. Heat losses occur via air moving through gaps in the building, or heat leaking through the fabric where there is no insulation, so in old, draughty, uninsulated homes, you end up using much more energy to stay warm. That's why it's a good idea to have insulation in your roof rather than leave it empty – or even worse, say, open up a giant spinning vortex to a Hell dimension that emanates from the top of your property. Your gas bills would be through the roof. As would you.

There are lots of extra benefits to decarbonising heat, such as having comfier homes, lots of jobs, better air inside our buildings, and, depending on how you do it – because you won't need one – no longer having your ears bleed when testing the carbon monoxide alarm. For the UK, it would also mean depending less on gas imports.

However, while there are upsides to decarbonisation, it is hard to undertake due to the upfront costs and the fact you are literally messing with people's homes. There have been calls for the UK government to ban new gas boilers by 2025, in the same way they have banned new petrol and diesel cars.[5] At present, government policy is only to ban new gas boilers in *new* homes by 2025. This should obviously already be happening.

To those who think it cannot be done, there are good examples. About half a million pounds was spent on improvements to Edinburgh Castle, and this reduced its energy use by 30 per cent and emissions by 40 per cent.[6] The money saved covered the costs within just five years, and they have now saved an extra £300,000. It clearly makes sense to do it: you just need the upfront cash. This is why government intervention must play a role for those who cannot afford it.

In hotter countries, they have the opposite problem: keeping their homes cool using technology. It is a little-known fact that the character in *Saved By the Bell*'s full name was actually Air Conditioning Slater. There are, however, quite a few similarities between the way we use energy in homes in the UK and, say, Kuwait. For instance, in the UK we leave our heating to come on during winter, while we are away, in order to stop the pipes freezing. In the same way, during summer in the Middle East, when people vacation to cooler climates, such as Coventry, they often leave their air conditioning on while they're gone to stop the plastic in their homes melting.

Air conditioners can contribute to global warming in two ways. Firstly, about a fifth of the electricity used in buildings around the world is, at present, for cooling.[7] Secondly, like fridges, air-conditioning units use GHGs called fluorocarbons, which, if they leak and escape, are extremely potent at warming the earth. Global demand for air conditioning is expected to triple by 2050.[8] Hopefully the electricity comes from more renewables, otherwise it means more fossils, more warming and more air conditioning, in what I assume the air-conditioning industry call a deadly bonanza.* Even in the UK, a decent fan is now an essential household item.

I had a quick look at the Energy Performance Certificate (EPC) that tells you about the efficiency of your home and found that our new home is an F. F for Fucking Fr**zing. F is the second worst. The rating goes all the way down to G. G is what I presume is given to dilapidated public toilets on the tops of windy mountains. Our house is one step removed from that. It could be slightly worse, but it could also be much, much better. What an estate agent would call having omni-directional potential. I didn't feel too bad until I looked up the EPC for Queen Elsa's Ice Palace from *Frozen* – and even her place has got an E, for Christ's sake. It's annoying, but I suppose I should just Let It Go.

Let there be light!

For once, let's start with one massive success story. Energy-efficient LED lighting uses about 75 per cent less energy than tradition methods, wasting less energy as heat, and it lasts twenty-five times longer while still brightening up our lives.[9] As a consequence, LEDs have helped to reduce emissions, and in 2017 they helped save the equivalent of around half a billion tonnes of CO_2 across the world. This is great news for emissions, but bad news for cartoonists, who might have

* Earlier, I said invest in umbrellas. You should also invest in air conditioning. A company that somehow does both will one day rule the world.

to think of a new way to show that someone has just had an idea.

And it's not only in our homes: LEDs are also saving businesses money and reducing their running costs. This is useful when you're in a twenty-four-hour ASDA at 3am in December and yet it is brighter than the summer solstice At first, LEDs were fairly crap, but improvements mean they are now pretty much indistinguishable from using other types of lighting. Other lighting manufacturers are incandescent about this. India saw a boom recently, going from 5 million LED bulbs used in 2014 to 670 million in 2018. Prices dropped from about £4.50 to £0.78 per bulb during that time, which has prompted a change in the phrase 'cheap as chips' to 'cheap as light-emitting diodes' in many Indian households. The popularity of LED bulbs in India has produced annual energy savings equivalent to powering the country of Denmark.[10]

One other method to save energy is switching lights off when you're not in the room. For some reason, I have a weird habit of always forgetting to switch the bathroom light off when I come out. Maybe I just never want the good times to end. Yet at other times, I will happily sit in a pitch dark living room, staring into the abyss, to save energy.

Insulation and efficiency

Many homes are like the operating system Windows – not particularly well-designed, but we just accept it and keep using it anyway. Colleagues of mine at University College London have created an energy efficiency map, the London Building Stock Model. It shows the whole city, with all the buildings scaled on how efficient they are.[11] Buckingham Palace scored highly, which surprised me as I've always found most people of the Queen's age turn the thermostat up full blast. But maybe when it gets cold, instead of turning the heating up, she just throws another corgi on the fire. King's Cross Station was poor, which didn't surprise me, I've never had a warm welcome arriving there. This map means anyone in London can go and check their home and see if it's better than their neighbours': a little healthy competition to help save the planet. You might frown upon snooping on your neighbours, but in London talking to them is frowned upon even more.

There are many steps we can take to make our homes more efficient, even if you're not planning on getting a new boiler anytime soon. For example, you can simply turn down the thermostat ever so slightly. I am already feeling the dad DNA coursing through my veins. Recently, I've started tinkering with the thermostat, moving it between various 0.1°C increments to get the right balance around the house. Being this tedious only gets you so far, though. Smart heating systems can help you be less pedantic.

Better insulation is needed in old homes like mine so that we don't require as much energy to heat them. Wall and attic insulation can help (this is the house equivalent of putting on an extra jumper). In the front room, where I am typing this in the middle of winter, cracks are endemic, and it is absolutely Baltic. There is a single-glazed glass panel above the front door, the letterbox doesn't shut properly, and there are spaces around the door frame. I have neither the time nor the knowledge to fix it, so I am sitting with a blanket on my lap like an old man in a care home. Dr Faye Wade (more on her later) advised me to put up heavy curtains in front of the door. Well, she told me to get a new door first, but that seems too much for today. Windows can also help. Double- or even triple-glazed ones, I mean, again not the awful operating system.

How do we get people on board with all this retrofitting? It isn't the sexiest topic. Basically, I think we need to *Queer Eye* our houses. Get a bunch of fabulous gay guys in retro outfits to retrofit our homes and make our lives better. Except the makeover shouldn't result in an emotional journey – unless you cry tears of joy at cheaper energy bills. Which, personally, I do. It might take the Fab Five too much time to go round millions of homes one by one, so maybe lookalikes are the way forward. Or we could get the Fab Five to convince sceptics. Queer eye for the climate denier guy. I don't know.

All I do know is we need to make it sexy and entertaining, because we need to speed up the introduction of energy efficiency measures. To meet our UK climate targets, it has been recommended we need to insulate 545,000 lofts, 200,000 cavity walls and 90,000 solid walls each year. In 2019, we only achieved 5 per cent, 21 per cent and 12 per cent of these three targets, respectively.[12] The Kardashians have finished their long-running reality series, but maybe they could relaunch it as *Keeping Up With the Kardashians' Plans for Loft Insulation*. I mean, if anyone is up for having any sort of work done, it's Kim Kardashian. If she could do for cavity insulation what she did for Botox, I really think it could turn things around.

The cornerstone of super-efficient buildings is the *Passivhaus* concept – a German design where buildings need almost no heating or cooling whatsoever. (This is not to be confused with a passive-aggressive house – the type of house I grew up in.) By using these standards, Brussels has gone from having one of the most inefficient housing stocks to becoming a shining example of efficient buildings.[13] Someone calculated how many cats you would need to keep in a passive house in order to heat it: 17.35.[14] Though really, you'd need eighteen for it to work. And, given the whole 'cats killing birds' issue we had earlier in the chapter on renewables, maybe this should be the cats' punishment – put to work to heat our homes? Hamster wheels for cats, running endlessly till they collapse with exhaustion. Financially, it turns out using felines to heat your passive home would be about twenty times more costly than using gas. But boy, will you get some social media content out of it.*

One major problem with making homes more energy efficient is landlords. When you rent, there is no incentive to improve the home you're living in: you don't own it. And landlords aren't exactly proactive when it comes to undertaking work that would reduce the bills of their tenants. This is what we call a market failure, which is a polite way of saying that, currently, nobody gives enough of a crap to do anything. Therefore, there is a major need for government intervention in the rental market to ensure low-carbon, climate-resilient housing stock. Governments need to start growing some balls in this area. Alternatively, we could just ban landlords. This is why I have always been sceptical of using the messaging 'We only have one home' when talking about climate change. The people with the highest emissions usually have two homes, they don't understand the concept of one home. Most young people cannot get on the property ladder as it is. Millennials are always being accused of spending their savings on avocados on toast. But do you know why they do that? It's because in

* Don't try this with dogs: they smell more and you lose more heat from opening windows.

lots of these hipster restaurants, the meals come served on slates, and if they can collect enough of them, they will be able to start building their own property. Find somewhere that does eggs Benedict on a fuse box, and you're laughing.

Heat (not the film)

We can take the numbers I mentioned earlier to do a quick calculation and provide a rough estimate of the scale of the heat decarbonisation problem in the UK. If 85 per cent of 28 million buildings use fossil gas, then that's about 23 million boilers that need changing by 2050. From the start of 2022, we have 1,456 weeks to do that. I make that roughly 15,800 gas boilers a week that need replacing with low-carbon alternatives over the next twenty-eight years. I'd love to help out with trying to achieve this, but I've already spoken about my competence in this area. What we need is an army of Joels, I mean Dads. And we are still, as I write this, building houses with boilers attached to the gas network, and with oil heating. This is the height of stupidity and either the construction and gas industries haven't quite grasped the scale of the change yet – or they have and are eking out the last drops from a dying business. Whichever, a massive shift is coming.

What should we use to provide our heating? I asked the coolest heat expert I know: Dr Richard Lowes, previously an academic from Exeter University and now at NGO The Regulatory Assistance Project, who also runs an Instagram account called Heatpunks and has been wonderfully described by a civil servant as 'the bad boy of UK heat decarbonisation'. The good news, he tells me, is that countries in lower latitudes only need to heat water: 'From North Africa to northern Spain, is fairly easy to deal with. You don't need space heating, you just need hot water heating. And so the technology is quite easy. You've got more sun, you've got even cheaper solar than we've got here. It's just about scaling them up and regulating them.' Those of us in the higher latitudes, though, have greater heating demands.

Heat pumps appear to be the cheapest approach to decarbonising our heating in the UK.[15] They are essentially the opposite of a fridge and sort of look like air-conditioning units that sit outside your house.* Heat pumps make heat from the ground, the air or water. The main reason that heat pumps are a major solution is that they already work at scale and would immediately reduce UK heat emissions by a factor of three – and this would fall even further as the electricity grid gets greener. However, they are pretty expensive to install at present, and at the moment the UK only installs around 30,000 heats pumps a year, compared to 1.6 million gas boilers. The Committee on Climate Change (CCC) advises that to meet our net-zero targets, 1 million heat pumps a year will need to be installed by 2030.[16†] The

* You might think that a cooker is the opposite of a fridge, but that's actually its natural enemy, which is a different thing.

† The CCC is an independent body that advises the UK Government on climate matters. They are actually called the Climate Change Committee now and not the Committee on Climate Change. They changed their name in 2020 because everyone has always got it wrong. Not me. A significant chunk of my PhD was on the CCC, and I have gone back through and found that I refer to the Committee on Climate Change thirty-nine times in over three hundred pages, and therefore I refuse to stop calling them by that name otherwise it was all for nothing.

CCC estimate that 10 million homes are suitable for heat pumps, and that another 10 million could be made suitable. The UK government has committed to installing 600,000 heat pumps a year by 2028 – which is one of those years that still sounds deceptively futuristic, but is basically next Wednesday, and if you can find a plumber today, that's when they will turn up to install anything.

We probably need a ban on new boilers combined with subsidies for heat pumps to get this moving now. Heat pumps have been used in Sweden since the 1980s, alongside district heating. District heating is the best solution for dense housing. This is where central heat sources in towns are distributed around neighbourhoods. Denmark has been big on district heating for some time. It would be possible to introduce district heating in about a fifth of UK housing. Local government is important here, because heat will be solved at a localised level.

Hydrogen is another option for heating, and can perhaps be seen as a less disruptive way of continuing with the current system, i.e. replacing the existing gas grid with green hydrogen made from renewable electricity. Hydrogen is being tested in 300 homes in Fife. However, when you dig deeper, it becomes clear that there are significant issues with hydrogen heating as a main route. It is considerably less efficient and green hydrogen would require four times the amount of electricity as heat pumps.[17] 'If we didn't have an existing gas infrastructure, they wouldn't even be thinking about doing it,' says Professor Tadj Oreszczyn, founding director of the UCL Energy Institute. Hydrogen is good for anything you cannot electrify, so it's great for primary steel-making and high temperature heat for industrial processes, as well as being a storage option for excess renewables – but this is not the case for low-temperature household heating. That's why heat pumps using green electricity make the most sense. Heat pumps also have the advantage that some systems can double up to be used for cooling, too. In the same way that in hotels you get hot and cold air, the same could be done in our homes – except, hopefully, without us having to spending twenty minutes trying to work out the controls before finally giving in and calling reception for help.

Now, all of this sounds like a lot of expensive work, but if we don't do it now, we'll be having to retrofit all our homes anyway in a much warmer world to keep them liveable. We have done it before. In the 1970s we transitioned heating in the UK from 'town gas' to 'natural gas' after the discovery of significant North Sea reserves. Town gas sounds like a nice, old-fashioned name, but, like most old stuff, it was incredibly dirty and made from coal. Besides, 'a lot of work' is actually what we could all do with right now, as our economy recovers after a recession-inducing pandemic. By 2030, it's estimated that there will be 160,000 jobs in low-carbon heat in England, and another 145,000 jobs in services such as lighting, insulation and control systems. There will also be at least one job in bleeding radiators, probably for a man named Joel.

The question is whether we have the skills to undertake such a massive task. I asked Dr Faye Wade, a built environment researcher I know at Edinburgh University, whether we had enough people to do the work that needs doing over the next decade? Her answer was, quite simply, 'No.' I wish I hadn't gone all the way to Edinburgh.

Cooking and fires

If we remove houses from the gas grid, then it is not just heating that will have to turn electric – cooking will, too. Cookers have, in the past, either used gas or electric coils. And people have tended to gravitate towards the fossil option due to a combination of extremely good marketing of pictures of little blue flames, together with our evolutionary pyromaniac tendencies. But cooking with gas inside your house is an idiotic idea of biblical proportions. Burning a fossil fuel directly in your kitchen creates nitrogen dioxide and other indoor air pollution. Which is terrible for, I don't know, small babies. And we have a gas cooker in the new house, with no extractor fan having to open the kitchen window ever time we cook, even in winter, and now I'm worried I'm a terrible parent. However, the new kid on the block is much better: induction hobs. These are the way forward, as they are super-efficient, and you aren't breathing in unnecessary indoor air pollution. I just need to sell enough copies of this book to buy one.

And talking of indoor air pollution. There's a beautiful little wood-burning stove in the front room of our new house. I was so excited to use it, to pretend that I'm a character in a Jane Austen novel coming home from an afternoon of chopping wood to set the fire for my family of an evening. In fact, pretty much all the terraced houses on our quaint little street seem to have a wood burner; you can smell them coming up the road. They're all old buildings, probably Edwardian, and this old-timey cottage vibe was one of the things that we found enticing about the house. I clearly wasn't thinking with my environmental hat on when we were looking at property (like

Indiana Jones' hat, it's hard to keep on all the time). Turns out, having wood-burning stoves that produce soot in our homes is a bad idea. We are just used to it as a hangover from a bygone age where an open coal fire was the norm. Even I grew up with one, and I admit it was lovely, but I will also probably die younger than I should – so, swings and roundabouts, eh? And so we've decided that our wood burner will only be used once or twice a year on special freezing occasions, if at all.

I bled the radiators with the tiny, cute key. Once the radiator was back on the wall, it had the knock-on effect of lowering the water pressure. I then spent the next while learning how to re-pressurise our boiler for the first time since, well ever. It's one of those boilers where you set the timer using tiny black teeth on a dial that you flick up and down. It's amazing how, in so many parts of human society, there is endless innovation, and in others we just go 'sod it, lets use tiny black teeth'. Anyway, I did it: I got the heating and hot water back on, the radiators were silent, and I felt like maybe I would make it as a dad after all. I texted Joel to tell him the good news.

20

Should I go vegan? I'm not sure I want to go vegan! Do I have to go vegan?

'I don't like last bites' was apparently my catchphrase as a toddler, says my mother. I'd always refuse to eat the last part of any meal. Little did I know, as a tiny little blond Scottish boy, that my steadfast refusal to finish a meal was contributing to climate change.

'Food – it keeps you alive.' That should be the slogan. Like most people in the world, I love edible things. Stuffing my face is obviously the best part, but I also enjoy cooking, as well as a stroll around a supermarket, picking up a punnet of blueberries, looking at the price, laughing out loud, then throwing my shopping basket at a wall and heading off to steal my own blueberries from a local bush.

But our coffee, our oranges and even our fish suppers are all under threat from climate change. Even wine is under threat.

With 2°C of warming, 56 per cent of the world's wine-growing areas may no longer be suitable for growing wine.[1]

Yet our diets will also play a crucial role in trying to solve climate change.

The whole global food supply chain is responsible for around a quarter of all worldwide greenhouse gas emissions (GHGs) created by humans.[2] This is something that we have only recently begun to understand properly, as it's rather complicated. We need to feed a hungry global population that has increased from 2 billion to almost 8 billion people in the blink of an eye, and is expected to reach almost 10 billion by 2050.[3] How the global population is fed is, therefore, important. We need to make sure everyone is healthy while causing minimal impact on the planet. Plus, diets have been changing too. Economic growth is bringing families out of poverty and changing what people prefer to buy, resulting in the increased consumption of sirloin steaks and Haribo Tangfastics. It's a toughie. It is estimated that between the years 2000 and 2050, we roughly need to produce as much food as has ever been produced in the history of mankind.[4] And I'm not going to sugar-coat that fact for you, because it would require more deforestation to grow the canes.

Waste rot, want rot

Food-wise, the easiest thing you can possibly do to combat climate change is to simply waste less food. If food waste was a country, it would be the third-largest emitting country in the world after China and the USA.[5] It would also almost certainly be the smelliest. About 30 per cent of the food produced worldwide is wasted, and this is estimated to cause about 8 per cent of global GHGs.[6] That is enormous. Therefore, only buying what we need is a huge part of the solution – plus it saves you money and time. Winning's win–win–win guarantee.

It's also easier said than done. I know I'm terrible at letting bananas blacken beyond the point of no return in my fruit bowl. Personally, I believe the best invention to stop us wasting food is already with us: the food bin. I eat everything, regardless of when it expired, just so I don't have to go near the disgusting mutant food bin. One look inside it, and you will never want to waste food again. Inside, when it's full, it looks like the film *The Fly*, except instead of Jeff Goldblum, it's a rotting courgette. And the stench is beyond anything the human nose is supposed to inhale. My wife once spilled it on herself while she was taking it down the stairs, and she had to be put in quarantine for a month.

The food waste from your bin is useful because many local councils burn it for electricity. Turning waste into fuel sounds great, but what happens if one day we have an incident – some sort of bin-juice Chernobyl – and we need helicopters to drop Febreze on to the power plant's core to get rid of the smell? If you don't have a food bin, then composting is a good way to reduce food waste emissions rather than just letting it go to landfill. While utilising our waste is good, it is better still to waste less. We can do this by planning ahead. If you have too much food, try using your freezer more or being creative with leftovers. Easter egg lasagne is going down an absolute storm in my house.

If you often eat at restaurants, it can be harder to minimise food waste, as you don't have much control over portion sizes. You can take away your leftovers using 'doggy bags', but this requires you to say the word 'doggy' to a stranger, and that always makes me feel rather uncomfortable. We need restaurants to get on board with minimising food waste, which will also save them money. It is essential to get away from the idea that bigger is better when eating out. Quality over quantity is most important when it comes to food. In the USA, portion sizes tend to be massive, with steaks the size of an SUV. That needs to change, both for environmental reasons and to minimise unnecessary calories.

It gets even more complicated when social traditions around food are considered, as it can be polite in some cultures to leave food on the plate to show you have been generously fed. Poverty can exacerbate issues too. In developing parts of the world, a lack of refrigeration or poor infrastructure means food often goes off before it reaches the end customer.

I've started using apps such as Too Good To Go, where you can cheaply buy food that would otherwise be thrown out from establishments near you, and Olio, which lets you share food. The good news is that the UK has reduced its food waste per person by 7 per cent in the last three years, which is enough to fill the Royal Albert Hall ten times.[7] (And let's face it, that would be far more fun than the proms.) It's totally possible to make these changes: Denmark has reduced the food it throws away by a quarter in five years.[8]

It's not just the individual's responsibility. Households are only responsible for about half of food waste in Europe.[9] The whole food chain is important too. A third of European fruit and veg are rejected due to being misshapen or for other purely aesthetic reasons.[10] Stop body-shaming broccoli.

Food glorious food

Of course, what food you decide to put in your mouth also matters. Switching away from meat consumption towards a more plant-based diet can have a big effect. The EAT Lancet Commission, an expert academic group, have recommended a combination of foods that would be both the healthiest for humans and most sustainable for the earth. It's called the planetary health diet.[11] And no, it's not a Mars a day. Wrong planet. The study suggests this could feed 10 billion people while staying within the planetary boundaries. What is it? Well, it is remarkably similar to Mediterranean and Japanese diets, which makes sense, as these are the places we associate with people tending to live long, healthy lives. The diet requires about half your average plate to be made up of fruit and vegetables. Yum. The rest is whole grains, plant-sourced proteins, such as nuts and legumes, as well as unsaturated plant oils, a very small amount of dairy and meat, and starchy vegetables.* So, less meat and potatoes and more beet and tomatoes.†

I gave being vegan a shot. I did meat-free Mondays, I took her for some tofu on Tuesday, we were making Quorn by Wednesday, and on Thursday and Friday and Saturday, with a nut roast

---

* Legumes, I believe, is a fancy name for beans, though Heinz Baked Legumes doesn't quite have the same ring to it.

† This joke only works if you say it in an American accent. Also, tomatoes are technically a fruit not a non-starchy vegetable, but peppers doesn't rhyme with potatoes. If the footnote is longer than the joke it is probably not a good joke. I am leaving it here, however, so you understand the lengthy, tortuous writing process I go through.

on Sunday.* I genuinely did find it tough, though. My cooking repertoire is mostly fish pasta and pasta with fish. Plus, I like a burger as much as the next Five Guys. So, it didn't last. But I have reduced my meat and dairy consumption substantially. Now I probably have fish once a week, and meat maybe once every two or three weeks instead of two or three times a week. My wife does a mean teriyaki tofu dish, but I need to up my vegetable dishes repertoire beyond a simple stir-fry with cashew nuts. There are a lot of plant-based alternatives out there now. At barbecues, I'll have Beyond Meat burgers and Linda McCartney sausages. Lab-grown meat is the next step – it will taste like the real thing, because it essentially *is* the real thing.[12] Insect-based foods are another potential solution; eating moths will be a way of taking revenge for all the jumpers of mine they've ruined.

The two weeks directly after the birth of our son weren't too bad, as my mother-in-law stayed with us and she's a pescatarian, so we ate pretty well. But once she left, all that quickly went out the window. For the next few weeks, we mostly survived on frozen meals made the month before, takeaways, and scraps from our neighbours, as Joel popped by with leftover barbecue food. Slowly but surely, we have improved, and meal plans for the week ahead have helped us reduce the amount we're buying and wasting.

---

* I have no idea if Craig David is vegan.

Meat your maker

Eating meat is bad for the climate for a couple of reasons. One is that, as a by-product of their digestive system, ruminant animals (i.e. ones who have several stomachs, like cows, sheep and deer) burp methane which is a harmful greenhouse gas.* As mentioned earlier, methane is about 80 times stronger than carbon dioxide at trapping heat in the first 20 years after it is released – which is bad – although it doesn't last as long in the atmosphere – which is good.[13]

The huge mass production of animals for food is unsustainable at present. That said, some groups have falsely overstated the impact of livestock. The documentary film *Cowspiracy* claimed the industry is responsible for 51 per cent of the world's GHGs.[14] A non-fact that comes from a non-peer-reviewed report. The real number is around 16 per cent, which is still absolutely large enough to require serious attention.[15] The same producer recently made *Seaspiracy*, a documentary about issues with the fishing industry. Many have pointed out that this should really have been called *ConspiraSea*. I refuse to watch it, as not bothering with an excellent pun is far worse than anything the film could possibly reveal.

Another strike against meat is that animals are often fed plants that could simply be used to feed humans instead. It takes a lot more energy to grow lots of plants to feed an animal

* Nobody tells you that writing a book means sitting on your arse for an entire year – ending up with four stomachs because you eat a vegan almond Magnum every single night to keep your spirits up.

that I'm going to eventually chow down on, rather than just sticking those plants directly into my mouth. Farm animals are often fed soy imported from Latin America, which can contribute to deforestation as trees are cleared to make way for new agricultural land. Since the 1970s, it is estimated that a fifth of the Amazon has been lost.[16] And I don't mean down the back of a couch. I mean razed to the ground. A recent study estimated that about a fifth of the soy coming to the EU from Brazil's Amazon and Cerrado regions comes from illegally deforested land.[17]

What if I don't want to give up meat entirely?

Some meats are also far lower emission than others. Unfortunately, those tasty moo-machines are the worst option at present, as they are big burpers. However, there is a bit more to it. Grass-fed cows are usually better than cows reared on industrial farms. Firstly, because often the grassland can't be used for growing crops anyway. In fact, some studies have shown that grass-feeding can help local soils to absorb carbon, as mentioned in chapter 13. Secondly, grass-fed beef avoids the significant run-off of massive heaps of cow manure from industrial cattle farms that produce methane and often ends up in local rivers. However, to meet (not a pun) our current beef demand using only grass-fed livestock would be extremely difficult, even if we used all the unproductive land available. And these cows still produce methane. Some reduction in beef consumption is probably necessary.

In the future, feeding cows seaweed might be one solution, as studies have shown that this can reduce their methane emissions.[18] But it certainly isn't happening yet, and I doubt you'll find bovines chowing down on your local beach anytime soon. Cut down on red meat and try to eat high-quality, grass-fed local animals when you do eat it.

What about other delicious animals? Well, while chickens require feed, their impact is smaller. Eating a fish-based diet has a lower emissions impact than consuming a standard meat-based one,[19] but there are a number of other sustainability issues with fishing that I don't know about because I refuse to watch that stupidly named film.

Do I have to go vegan, then?

Well, there's not a huge difference between being vegan and being vegetarian from an emissions point of view, although the former is slightly better. Cutting back on meat is the first step, and the most important. Start with reducing red meat intake, then try cutting back on poultry and fish, and finally try going veggie or vegan if you can. Many sports stars have started going vegan, including Venus Williams, Lewis Hamilton and Hector Bellerin. But if it's not for you, that's OK too. For most people, it is a journey rather than an overnight change. Fewer young people are eating meat nowadays, although this may be less down to environmental reasons and more to do with them growing up watching too much *Peppa Pig*.

There are also extra health benefits to be gained from reducing your meat consumption, such as lowering the rate of heart disease.[20] It is estimated that a vegan diet could avoid 8 million premature deaths a year globally and reduce health spending by about $1 billion a year[21] – which will buy you three or maybe even four quinoa bakes from Whole Foods.

I drink a lot of coffee as an academic, so I investigated the emissions associated with a cup of joe. I found out that if you have a latte (because you're a big baby and you need a big glass of warm milk like a big sleepy toddler), then your emissions are about double that of an espresso and 62 per cent higher than a flat white.[22*] Espresso is, therefore, the best option for the environment – as if Italians needed another reason to be smug.[†] The main permanent difference I've made to my diet is reducing my dairy consumption. My wife is allergic to dairy and has drunk soy milk all her life, which has about three times less emissions than cows' milk.[23] I tried soy, but was never a fan. Now, I only use oat milk – which is especially strange when I have it in my porridge. You know, oat-on-oat action. It makes me feel like I am the most Scottish man in the world. As if I love oats so much that I want to ground down oats into liquid so I can pour them over my oats. I LOVE OATS! Plus, it is really good environmentally.[24]

Passports for parsnips

Another climate-beneficial choice you can make is to eat fruits and vegetables that are in season. Eating locally tends to be best, especially if the foods are not grown in a greenhouse. If you're worried about the air miles of your Asda shop, then

---

* I miss lattes.

† This chapter should perhaps have simply provided a two-word answer on how to reduce your food and drink emissions: 'Become Italian'.

eating fruit and vegetables that don't need to be flown in from another country is more climate-friendly. In other words, don't buy tomatoes that have their own frequent-flyer points. In some cases, though, it's actually better to eat fruit and vegetables imported from places where they are grown more efficiently when they arrive via ship or rail. But to prevent complicating things, if in doubt try to eat from local suppliers or grow your own, you mad hippie.

Changes in the broader food system are also needed though. We need to change agricultural practices and make low-carbon fertilisers that can be produced from electrolysis using renewables (current fertilisers are produced using fossil gas). In an ideal world, companies would have to display information about a food's carbon footprint on packaging so consumers can make better choices. Helpfully, some companies, including Quorn and global food conglomerate Unilever, are introducing just that.[25]

Are things changing? Well, France has been discussing reducing meat on menus by making vegetarian options the standard for public canteens and introducing one meat-free day a week in schools.[26] And if France can do it, by God, anywhere can. So overall, waste less food, eat less meat, and become xenophobic about peaches.

I will leave you with a vegan recipe that my mate Ian has made every day since his wife left.

TOAST UNDER BEANS

1. Open can of beans.
2. Tip beans in pan and turn on heat.
3. Scroll through pictures of ex on Facebook.
4. Put toast in toaster and turn on.
5. Find the last picture of the two of you together on Facebook. Stare at it. It was at Colin's party. She seemed so happy, though. You remember that you bought her that dress; it suited her. You stare at her eyes to see if you can tell whether she'd already decided by then. The toast pops up and startles you back to reality.
6. Pour beans on a plate.
7. Place toast slice on top of beans.
8. Put second plate on top and turn whole thing over.
9. You now have toast under beans. Plus, twice the amount of washing-up, which makes it feel like she never left.

21

Why do we have so much stuff?

Kids come with a lot of extra stuff. Obviously, not exactly when they arrive out of the womb – it's not like they've been ordering Amazon packages in there that they've been left with a neighbour. I mean that once they've arrived, they need an awful lot of bits and bobs.

Including bits and bobs for boobs. Until very recently, I did not know much about breast pumps. In fact, I knew nothing at all. I didn't know my membrane from my shield. If I'd chosen it as my specialist subject on *Mastermind* I'd have got zero – and probably a few odd looks from John Humphrys. Thankfully, my friend Andy works at an upmarket breast-pump company. He was able to procure us an unused second-hand pump from someone who was too stupid to work out how to open a box. As Andy got into the nitty-gritty of the ways in which this model was superior to other competing booby-juicers, I enquired about the lifecycle emissions involved in the manufacturing process of a pump, and we got into a lengthy back and forth on WhatsApp. He said he'd come back to me with an answer. In retrospect, maybe I should have just said thank you.

Drowning in plastic

The stuff we buy, the gifts we give, our goods and products, our belongings, our gadgets and knick-knacks – they all have an impact on the planet, because making things requires resources and energy. We often just don't perceive it in our day-to-day lives. We want the shiny new product, not its life story (except when we're watching a Pixar film about anthropomorphic mobile phones).

When it comes to chucking all this stuff out, the systems are getting more and more complicated. First, there's the bin-bin. You know, the No-Man's-Land bin, for all the stuff that doesn't have its own specific separate bin. It used to be the only bin. *The* bin. Now it's the bin-bin. Then, obviously, there's the food bin, which, as I mentioned earlier, is important but toxic, and I will not spend another second thinking about that stinky nightmare.

And then there's the recycling. The song 'Yakety Yak' wouldn't get made nowadays – not because there's anything offensive in it,* but because the papers and the trash are now taken out on separate days. I spend ninety per cent of my life now doing the recycling. Every day is Groundhog Day. The pile builds up and up. I empty the recycling outside only to go back inside and it's there again, like one of those cartoons where someone takes a cat that they hate far away to a forest and then gets home to find its already there. Except that instead of cat, it's empty cartons of oat milk and nappy boxes. I recently forgot to put the recycling out before collection day. Our bins

---

* Unless you're a yak.

only get collected every two weeks. I was apoplectic, and genuinely had a small breakdown and cried. I mean, sure I had barely slept for more than three hours straight for months on end. But still. I shed tears about recycling.

Luckily, metal cans, as well as paper and cardboard, are highly recyclable. Shout out to my fans at the European Corrugated Board Industry, who follow me on Twitter.* We have started using the company Who Gives a Crap for all our loo roll needs. They use recycled paper and plastic-free packaging, plus they donate 50 per cent of their profits to building toilets for people that don't have access.†

The bad news, though, is that our foods and goods come covered in abundant plastic. Forty-two per cent of plastics made worldwide are used for packaging.[1] Plastics are rightly a huge public concern, more due to ocean pollution than for climate change per se. The fact remains that the sheer amount of plastic is staggering. In 2015, over 380 million tonnes of plastic were

---

* Cardboard. That's my target demographic. And the worst part about it is the last time I looked, they had about 1,000 more followers than me. That made me want to walk directly into the sea – which will become increasingly easy to do over the coming decades.

† Hopefully, once they've sorted that, they can solve the bathroom crisis here in the UK. What crisis, you say? Well, the fact that the doors outside bathrooms are becoming more and more like quizzes. When I tried to go to the loo in a pub recently, I had to choose between going into a goose or a trumpet. Why put confusing pictures on the outside? It is not the time to be quirky. I mean, it used to be trousers on one door and a dress on the other. Now it's just two suit jackets, but on one door the buttons go down one side and on the other they go down the other side. Themed restaurants are the worst. Like a fish restaurant will have an anchor and an octopus, and there'll be a queue of people outside them going, 'I don't know what this means – I'm going to piss myself.'

produced – the same weight as two thirds of the world's human population.[2] Imagine how much I'd end up crying if I had to take that much recycling out in one go. And that is just one year. Due to accumulation over the last several decades, the amount of plastic on the planet is approximately double the weight of all the animals.[3] Hence why there is a garbage patch in the Pacific three times the size of France.[4] And with far fewer cute bakeries.

All this plastic lasts for millennia. The packaging you just placed in the recycling bin might still be here to see if Busted were right about the Year 3000. Plastic is being found inside fish and birds in a sort of weird reverse Kinder Surprise egg, and even in unborn babies.[5] The new circle of life means buying a kid's meal decades ago containing a toy that you rediscover years later inside a Fillet-o-fish – though thankfully, since January 2021, McDonalds have removed plastic toys from Happy Meals in the UK and Ireland.[6]

Many plastics cannot be recycled using current techniques and most plastic ends up in landfill. Lots are then burned to produce energy from waste – in 2019, the UK's emissions from incineration were higher than from coal.[7] Also, most of those little symbols you find on plastic products are meaningless. I mean, technically, they tell you what type of plastic it is. But they are designed to look like recycling symbols and are intentionally used on items that aren't even recyclable. Why? To make you feel better. To make you feel like the world is a safe place. That everything is being taken care of. It isn't.

Sure, plastic does have many important benefits for specific purposes, that can far outweigh any harms, like for equipment in hospitals, safety helmets and keeping food fresh. But the new pair of scissors you bought online that were just delivered won't 'go off' anytime soon – so why are they encased in plastic?

The skyrocketing use of avoidable single-use plastic is, of course, intentional. In the early 1960s, the plastic industry dreamed up this future hellscape. This quote is taken directly from the 1963 National Plastics Conference in Chicago: '[T]he package that is used once and thrown away like a tin can or a paper carton represents not a one-shot market for a few thousand units, but an everyday recurring market measured by the *billions* of units. Your future in packaging . . . does indeed lie in the trash can.'[8]

Get in the bin. A lot of plastics made by the petrochemical industry are simply new ways for companies to make money using by-products from fossil fuel extraction. If I thought I could sell my own 'by-products' to people despite there being no demand, I'd be pretty happy to.*

There are some things that will help. Package-free shops have been springing up, and some supermarkets are trialling this sort of thing too. Recycling organisations such as TerraCycle are great for all the extra stuff that can't go in normal recycling. It is also possible to undertake chemical recycling of plastics, which breaks them down into their basic chemical components. This allows new plastics to be made from old plastics (instead of using more fossil fuels) - closing the circular loop on plastic production, although little of this recycling is undertaken at present as it is too costly and there are little in the way of incentives. And there are now ways of making packaging from biodegradable material such as corn starch.[9]

People think because I work in climate change that I must care about recycling and know all about which items can be recycled. But I cannot stress enough how little of a shit I give about recycling.

---

* I imagine there is somewhere on the internet.

Even in my own house, I am asked whether this bit of plastic thingamy or that metal do-da should go in the recycling bin. I have no idea. I just make it up. I don't think I've once known the actual answer. It's nice that people engage with it, and it's best not to put them off by telling them that recycling is not as big a deal as people think it is. It cuts maybe about 3 per cent of the average British person's carbon footprint.[10] Doing all your recycling for five years is the same as taking a return flight from London to New York. So, by all means, do it – but maybe focus more on the big-ticket items. Plus, *you* aren't actually recycling anything. You are just undertaking, for free, the first stage in a collection process for materials that you never asked for.

All this is a long-winded way of saying that I do not care what you do. The idea that the onus is on individuals to do their bit and recycle has been another great public relations swindle.* The plastic industry wants you to buy into recycling, as it allows them to keep doing what they're doing. If you think that something is *being done* already, you won't push for more regulation to force them to change their practices.

I do care what companies, industries and governments do. The onus should be put on plastic manufacturers to take responsibility for the whole lifecycle of their products, or for companies further down the supply chains to start eschewing having their products wrapped in unnecessary nonsense. *You* did not buy the plastic packaging; it should not be your responsibility. (Unless you're one of those people that buys *Star Wars* figures and keeps them in the original packaging, then you're the most evil bastard out there.) Authorities need to legislate

---

* You may now be noticing that great PR is a remarkably consistent trend among companies entangled in the world of fossil fuels.

to ensure industries are responsible for their own waste. As individuals, the main thing we can do is to tell the companies that it is no longer acceptable, and we want them to change.

Now on to our final bin – the poo bin. Yes, nappies. The dilemma of whether to use disposable or reusable material to capture the endless stream of brown that emanates from their tiny bums. Is there a right answer? A single baby uses about 5,000 disposable nappies before being potty-trained – but reusables require washing at a high temperature. A 2008 UK government study found disposables had slightly lower emissions, based on a 60°C wash and one in every four washes being dried in a tumble drier.[11] Although machine efficiency improvements and a greener grid over the last thirteen years may have improved things for reusables. If you always dry on a line or rack, then reusables are likely best, especially if used on a second baby.* We have used a mixture of disposable and reusable nappies. It's a conundrum. You're damned if you doo-doo . . . it's a bit like deciding whether to dry your hands with a paper towel or a hand dryer.†

Catwalk carbon

Nappies are our first foray into fashion, but far from our last. Some studies suggest clothes are responsible for about 8–10 per cent of global emissions, although this appears inflated, as it double-counts chunks of agriculture, chemicals, electricity and

* We are not even *thinking* about another baby yet.

† The answer is apparently neither, but instead to dry them on your clothes, as this requires no energy whatsoever, but I was always taught this is disgusting.

transport, and the fact that trendy, high street designer clothes shops have their doors wide open in December. Two to three per cent seems more accurate.[12] Turns out flying is not the only thing involving a runway that is terrible for the planet.

The clothes industry is huge. At one end, you have Jean Paul Gaultier telling the rich they need to wear a new beret each season. And at the other end, is the rise of cheap, fast fashion. Anything that is mass-produced and quickly thrown away is, of course, a calamitous shitfest, as we've seen with plastics.

There are options to counter this. One tip to make your jeans last longer and save on washing, is to pop them in the freezer. Just remember to take your debit card out first, otherwise your bank account will be frozen. Obviously second-hand clothes and charity shops are brilliant. Improving the quality and variety of these would help more use them. Plus, websites like Loopster allow online buying and selling of second-hand clothes. Most of Oscar's baby clothes have been second-hand, passed down by other parents desperate to clear space in their attics. I think this is the way forward. There have been examples of climate-conscious teenagers beginning to regularly swap clothes with each other and using 'slow fashion' apps such as Nuw and Swopped that match you to someone of your size with a decent dress sense. Some brands, such as Gucci, Levi's, Patagonia and even Asda, are launching resale platforms.

The biggest impact of jeans and T-shirts comes from their production. Except for jeans in the USA, where over-washing and machine-drying make the ownership part worse.[13] Washing clothes less, reducing the temperature on the washing machine, or even washing them in cold water can help save energy – and it also helps clothes last longer. I'm seriously considering going for a cold shower wearing all my clothes, because (a) it will save emissions, (b) the washing machine is constantly on for all my son's clothes so I can never get mine in, and (c) it will make me feel alive. We also have a clothes line, which, in the summer, uses that big free orange ball of fire in the sky to dry clothes. During the winter I just blow on them. Turns out not having the space for a tumble drier is good news for the planet. According to the book *How Bad Are Bananas*, a 40°C wash generates about 590g of CO_2e, but using a tumble drier takes that up to 2,000g CO_2e in total.[14, *]

Good white goods

White goods are household items we do not purchase often, but when we do, our choices are important. Moving house, we had to buy all of them. The worst part is waiting for them to arrive: 'Thank you for ordering your new refrigerator, Dr Winning. It will be delivered to your house sometime between 8am and June.'

Fridges are especially hard-working as, like most comedians, they're always on. And as well as consuming all that electricity, they also require refrigerants, which, if they leak, are massively harmful to the atmosphere, and so must be disposed of safely. Plus, they're much like cars in that, while their efficiency has improved, they keep getting unnecessarily larger. A big fridge was always a fantasy of mine as a child. Filled with three cartons of Sunny D, Florida-style.

We now own a dishwasher for the first time, which has been an absolute revelation. It does the washing-up in half the time it took me to clean one bowl and a spoon. If you turn it on only

* Kids toy idea: Mr Tumble's Tumble Drier.

when it's full, usually dishwashers lead to lower emissions than washing by hand, even when you include the manufacturing emissions. But please make sure it is as full as you can get it. I simply refused to visit Ian and his ex-wife at their house because of how they stacked their dishwasher. They put it on after every meal. I was round one day and he had toast under beans and then proceeded to put the dishwasher on. There was literally one pot, one plate, one wooden spoon and one fork in the entire dishwasher. I refused to go back.*

Then there is all the other stuff we buy. Having a child has only added to this. Prams, sterilisers, bottles, red nightlights that make your child look like the devil. Children's toys are often made from plastic. I remember being so excited about toys in cereal boxes, and now realise that they are the worst kind of throwaway nonsense. I am not some mad environmentalist who would suggest simply giving your child a stick to play with called Sticky.† There's a middle ground. BuyMeOnce is a company that rates and sells products that are designed to last a lifetime, including baby clothing and accersories. My wife discovered a toy subscription service called Whirli, which we now use for all his toys. This means you can send it back if they're not interested in it or get too big, plus there aren't masses of toys lying around. Plus, less emissions. His favourite toy at the moment is a blue monkey, which we have called Lisa because he keeps trying to tear it apart.‡ As he chews on Lisa constantly and sleeps very little, I have had the Daft Punk song

---

* No dishwasher for him now.

† However, I am absolutely using cardboard boxes for him to play in – so I don't have to recycle them.

‡ You have to be a fan of the film *The Room* to understand that one.

'Get Lucky' in my head for several weeks now, but at the chorus I've changed the lyrics to 'I'm up all night to eat monkeys'.

A major problem is how manufacturers seem to plan obsolescence into their products so that we've come to expect goods to be replaced regularly.

Our grandparents' generation had the right values about not wasting anything. Sure, their views on gender, race and sexuality may have been questionable at times, but they were right all along about stubbornly using and fixing stuff until it is basically no longer recognisable. Not being wasteful is also a concept generally agreed upon by all sides of the political divide. 'Reduce, Reuse, Recycle' is the mantra. It has recently been rebranded as 'the circular economy', because to get people with money to listen, you have to talk about the economy, and also because circles are cool.

It is obviously preferable to reuse things. But what if your laptop breaks, just after your warranty has just bloody expired, due to your son puking up milk all over it, and you've got a first draft of your book due, and you don't know how to fix it? Luckily, there is another relevant 'R' to add in the fight against climate change. The 'right to repair' is being introduced at EU level, as well as in the UK and in some US states. This will make it a legal requirement that goods last longer and are easier to mend. For the EU, manufacturers will have to provide spare parts for ten years for larger household appliances. I realise that when people talk about rights, it's usually in reference to free speech or equality. I'm not sure Martin Luther King's speech would have had the same impact if his dream had been about being able to fiddle about with the drum of the tumble drier. But it is still important, as manufacturers will no longer be able to take the piss. Sweden already has a tax break that halves the cost of getting someone to repair your white goods.[15] Other helpful trends include campaigns like Recycle Your Electricals, and the rise of repair cafés. There's even a TV show called *The Repair Shop* on BBC One, but I haven't seen it – I'd need them to repair my TV first.

This book

What about books? Well, *Spot the Dog* hasn't got as many pages as this book, and so it is better for the planet – plus Oscar seems to prefer it. He only lasted for about a paragraph of this book before saying that it was alarmist left-wing propaganda and calling me a Trotskyist. For a long time, books relied solely on the paper industry, but this has changed recently. What has the bigger impact – physical books or an e-reader? Well, you would have to *not* buy about forty paper books to make the e-reader more sustainable.[16] Really, though, the best thing for societies – and the planet – are libraries. If you have borrowed this book from someone else, well done. Though I will be coming round to collect £16.99 RRP at your earliest convenience. Our bookshelf now has many, many second-hand books about babies, which we've tended to read in a panic at 4am because a noise sounding like a cat stepped on a hedgehog comes out of our baby while he's sleeping.

The interweb

Computing and the internet both cause emissions and help reduce them. Zoom and Skype calls help to reduce transport emissions from people travelling for meetings, and after growing in popularity during COVID-19, will probably become even more prevalent. Yet computers also rely on energy-hungry data centres around the world which use an estimated 1 per cent of global electricity.[17] It is estimated that the bitcoin industry uses more electricity than Belgium.[18] I thought bitcoin mining was just a term, but it does involve, at present, an awful lot of actual coal mining. Elon Musk recently distanced Tesla from bitcoin. Afterwards, *Bitcoin Magazine* tweeted 'When a climate alarmist demands for you to explain bitcoin's "alarmingly high energy usage," ask them what the future looks like without bitcoin?'[19] Not without the coral reefs or the Maldives. Without Bitcoin.

The best thing we can do is to keep pushing massive internet companies like Microsoft, Amazon, Google, Apple, Facebook, etc., encouraging them to be more open about their energy use and go 100 per cent green as soon as possible. You can also use the Ecosia search engine, with one tree planted for every forty-five searches you make. It is one of these B Corporations that balances profits and purpose – like Who Gives A Crap also is. After paying for costs and tax, it spends 61 per cent of its profits on planting trees, 16 per cent on green investments, 23 per cent on advertising and none on shareholders.[20] Using Google to find the name of the actor you recognise in the film you're watching is also a decent option, as it matches 100 per cent of its electricity consumption with renewables. In fact, Google's reporting is pretty detailed and open; for example, its disclosed emissions went from 3.3 million tonnes of CO_2e in 2017 to 15 million tonnes of CO_2e in 2018. This rise was actually a good thing because Google expanded its definition of the emissions it accounts for.[21] This should be standard practice among more companies. Microsoft aims to be 'carbon negative' by 2030, and says it will also offset all its emissions since 1975 by 2050. Facebook claims to already be net zero in its own operations and is targeting net zero across its entire supply chain by 2030. However, they also promoted adverts denying climate change 8 million times in the USA in the first six months of 2020.[22] Not sure if they have a target for that. Apple and Amazon also have pretty remarkable targets. Basically, if you have all the money in the world, it seems you can do whatever the hell you like.

Heavy metal

Manufacturing is a big piece of the emissions puzzle, and something we as individuals can do little about. Heavy industry should really have its own full chapter given its importance in causing climate change but it is almost impossible to write observational jokes about, so I'll keep it short.

The steel industry is responsible for around 7 per cent of global emissions, mostly because the main way, at present, of turning iron ore into steel involves burning lots of coal. Half of current global steel production takes place in China. To solve this, we need to go big on steel recycling which uses electricity – though I'm not sure which coloured bin skyscrapers are supposed to go in – and also green hydrogen looks a promising new technology for steel.

Cement is another heavy-polluting industry with a product that is used to make our buildings and roads. The way cement is made causes process emissions from the chemical reaction, which in turn requires carbon capture and storage to be significantly deployed across the world. I found it hard to come up with one solid joke about cement. It is certainly the hardest section to write. Most people from the industry (cementists?)

are stuck in their ways.* Fingers crossed I'll come up with something a bit more concrete for this bit eventually.

I received the following message from Andy about the fancy electric breast pump: 'Matt, you saved 25kg CO_2 by getting a second-hand pump.' That was good news. It was followed a minute later by another message. 'However, if you use it like the average American, with two kids, you will produce 14kg CO_2.'

I resolved then that we would not become American. I wasn't sure about the two kids though. I thought about the 14kg and then about the millions of tonnes of carbon created by companies. Then I got back to the job at hand: as there was a full box of recycling to take out.

* They're actually not: most cementists are very much on board with the net-zero transition and quite nice to boot.

PART 3

WILL WE CHANGE?

22
Three months

You are a badass – you can do this

As we approached Oscar's first Christmas, I found myself imagining all the Christmas Days he will see over his lifetime, wondering what our traditions will be, and whether, this week, his milk will taste like Brussels sprouts. I couldn't wait. I got us a real tree. Apparently, you need to use a plastic one for longer than twelve years if you are to offset it compared to getting a different real one each year. However, the best thing is to get a potted real tree that you can replant and use year after year. Maybe I'll do that next year. It hadn't occurred to me to do it before, as, until now, I'd not had a garden as an adult. There are also services that hire out trees and then take them back to replant, which is handy if you don't have a garden. Anyway, I learned about these too late, after I'd already lugged a small, fat tree back home.

It was the week before, exactly three months to the day after Oscar was born, that for the first time, the reality of the climate crisis properly collided with the responsibility of bringing a life into this world. Up until that point, climate change had barely crossed my mind: I'd been too busy sterilising bottles and changing nappies, and if it did enter my noggin then it was in an abstract, academic-type way that was devoid of emotion.

The previous three months had been a whirlwind of bonding, worrying and learning on the job. I quickly became Tom Selleck's character from *Three Men and a Baby*, mostly because I didn't have time to shave properly. If I'm honest, I'm probably more of a Danson. We slowly settled into a rhythm. Lazily spending time lying all together in bed during the early hours of the morning. Exploring new walks with him in the baby carrier and smelling his lovely wee head. Bath time was the best – he loved it. Trying to make him laugh because it was the single most wonderful sound we'd ever heard. Reading the same *Spot the Dog* book to him every night before bed, and trying to find new ways to add jeopardy to the fact Spot has lost Teddy.* And Oscar slept just enough most nights for us to be able to function. Not to function well. And not every night. But we can't complain. I mean, I can and do complain, but it could be worse.

And then it hit me. I was watching an online launch of the *Lancet Countdown on Health and Climate Change*. This year, for the first time, I was an author on this annual report, which is, unsurprisingly, about the important link between health and climate change. I didn't have to do anything except listen to the web stream, and so I decided to get up from the desk and move to the living room so I could sit on the sofa, on Zoom, listening to the report being presented through earphones while watching my new family.

* Also, my wife noticed that Spot only has two slippers. We've had lengthy conversations trying to understand why.

After a few minutes of introductions and such, it got into the meaty details.

'91 per cent of deaths from ambient air pollution occur in low-income and middle-income countries.'

And as I listened to these awful stats being rolled out in a serious tone, my wife was in front of me, helping my son to stand upright on his playmat that had a map of the earth on it.

'Premature deaths from ambient PM2.5 attributed to coal use are rapidly declining, from 440,000 deaths in 2015 to 390,000 deaths in 2018.'*

She was pretending to walk him across the world, and he was making grunting noises, i.e. he was loving it. I thought about babies across the world.

'However, total deaths from PM2.5 have increased slightly during this period, from 2.95 million deaths in 2015 to 3 million deaths in 2019. This highlights the need for more action.'

Something about the juxtaposition between what I was hearing and what I was seeing broke me, and I started crying. Not full-on sobbing: just a wave of emotion at how fragile life can be, followed by a tear or two. It passed as soon as it arrived.

OK, now, I know you're thinking: 'Matt, this isn't really comedy any more, where are you going with this?' Well, I'm just being honest about the fact that, sometimes, my life is weird. Working on these issues can be difficult, and finding the humour in it can be hard at times, especially now that I have so much to lose. A bit like NHS health workers do,

* Particulate matter describes particles in the air, like soot. PM2.5 is also known as fine particulate matter, with the number simply referring to the size of the particle. For above, essentially read it as air pollution.

as a climate-change researcher, you sometimes have to detach yourself from your emotions and see it as a job in order to get through it.

A child born today will face health impacts from the climate and human actions from the moment it arrives, and through its teenage years, adulthood and into old age. My son's entire life will be affected in some way by the decisions that have already been taken, and by what we choose to do now. It can be hard to comprehend all the babies that have been born over the last year and the greater difficulties they will face growing up unless more is done. But the thing is, Oscar will be one of the least adversely affected in the world. Already, families that are barely getting by, who are just trying to make ends meet in order to provide for themselves and their children, are having their circumstances altered by an unstable climate. From farmers in Uganda to the residents of Kiribati and the Saami herders in Sweden, it is becoming even tougher for the poorest people in the world to survive and look after their children. Climate change is making life harder for those whose lives are already the hardest.

A child born in one of the least developed countries will face much harsher consequences. Food production and food security are being threatened by downward trends in global yield potential for all major crops. Undernutrition can severely affect children's development in their vital early years. Children are also among the most susceptible to diarrhoeal disease, and experience the most severe effects of dengue fever. Nine of the ten most suitable years for the transmission of dengue fever on record have occurred since 2000.[1] Imagine one of these families, in the same position as us with a new baby, trying to fulfil their basic needs: being fed, watered, clothed, looked after and loved. All while the world becomes more uncertain. It seems impossible.

It's not all been me being miserable, though.

We did have a Christmas miracle. We saw Santa Claus. Actually, my wife saw Santa Claus out the window in the front room, being pulled by his reindeer up a lane next to our house. By the time I got to the window, I saw a bald man in a high-vis jacket. I told her she was going mad, that Santa wore a red suit and wasn't bald, and that she needed to get more sleep. She took the bait. We stopped running the bath for Oscar, and instead all three of us put on our warmest clothes, and went out on to the street to find Santa Claus. By the time we got out and up the lane, he was gone. I continued the banter about how she needed to stop making stuff like this up, and that just because we had a child now it didn't mean Santa Claus was real. Once around the block we went. We passed the house with the model of itself in the window. It had a Christmas tree both in the actual window and in the window of the tiny version. Then up a side street. No sign. We gave up and went home. Then, out of nowhere, he showed up again, right outside our house. I mean, it was a man far too young to be Santa being pulled in a sleigh by a Range Rover. We all waved and took some pictures. Apart from us, the street was empty. I WhatsApped our neighbours to find out what the hell was going on. They said it happens every year and was 'a weird charity thing', in a matter-of-fact way that suggested none of them were as excited as we were. Anyway, it was our first family Christmas memory.

23

What are we doing about it?

Well, firstly, it's worth quickly going over what we've decided, as the people on this planet, to do about the fact we're destroying the planet. How far away are we from ending humanity's contribution to climate change, and what needs to be done, and when, to get there?

International agreements

There is no 'technically' agreed level of what constitutes dangerous climate change. All warming is harmful to some extent, and thus a 'safe level of warming' is open to interpretation – and everyone and their grandmother has an opinion.*

The international consensus for some time has been that we need to hold the increase in global average surface temperature to below two degrees of warming above pre-industrial levels.

* My grandmother's opinion was that they should all listen to me, because 'you're such a nice boy, Matthew'.

Given that we've already warmed by just over 1°C, this is quite the challenge. Ironically, two degrees was also the upper level to which my parents were willing to support me through higher education.

The main method to get all countries to agree on a plan of action has been through the United Nations process. Each year, lots of delegates get together at the Conference of the Parties (COP) for a chinwag. The first one was in Bonn in 1995. There was a big one in 1997, when the Kyoto Protocol became the first global climate agreement, in 2009 in Copenhagen, where the failure to reach a new ambitious agreement was a major setback, and in 2015 where the Paris Agreement was struck.

Leo Hickman, editor of Carbon Brief website and veteran attendee of COPs, describes them as fascinating, immensely frustrating and utterly like *Groundhog Day* all at the same time. He adds that 'a lot of good does come out of them as well. And it creates a much-needed media moment, if I'm being a bit cynical, where the world collectively, even for a few days, is very focused on what world leaders are doing – or not doing – about climate change.'

After Copenhagen, everyone thought the UN climate-change process was dead. However, like a phoenix from the flames, the 2015 Paris Agreement took a new tack and was hailed as a truly successful global agreement on climate change. It achieved what many had failed to do before: it got everyone on board. And how did it do that? By letting countries do whatever the hell they want. Instead of trying to come together to get countries to agree on 'You do this, you do that', it basically just went, 'What do you think you can do?'

However, the end goal of the Paris Agreement is still ambitious.

It states that warming needs to be limited to 'well-below' 2°C and countries should pursue efforts to limit it to 1.5°C.* That half a degree's difference might not seem like much, but in practice it means saving many Pacific islands, which would disappear with warming above 1.5°C.† The percentage of global population at risk from severe heatwave is projected to be 14 per cent at a 1.5°C rise, but this increases to 37 per cent of the population at 2°C.[1] We can expect one ice-free Arctic every 100 summers at 1.5°C, but at 2°C this will increase to one in every ten summers.[2]

We must move rapidly. When Snoop Dog sang 'Drop It Like It's Hot', he was explicitly referring to worldwide greenhouse gas emissions in light of global heating. To stay below 2°C, we need to be cutting global emissions by around 3 per cent per year. And to stay below 1.5°C, far, far faster emissions cuts are needed: they would need to fall off a cliff of around an 8 per cent drop a year, every year, for decades.[3] A bit like how you've had a 7 per cent drop in optimism for every sentence since the start of this paragraph.

When zero sort of means zero

To end the human contribution to climate change, we need all countries to have, on average, zero emissions at some point in

* 'Pursue efforts' sounds like the sort of language I use when I tell people I'm planning on going to the gym far more regularly.

† This is a massive issue with climate change communication – small numbers can have massive impacts, but we hear them and think, 'That's a cute wee titchy number, how sweet.'

the second half of the century. This will get us to 'net-zero' emissions globally. The term 'net zero' is essentially a way of getting around the fact that emissions from some activities, such as agriculture and chemical processes, are too hard to escape. For any amount still being emitted each year, the same amount needs to be sucked out of the atmosphere.

The Paris targets have brought about a movement of countries committing to reach net-zero emissions. At the time of writing, countries covering 61 per cent of global emissions have net-zero targets that they've either adopted, announced, are considering, or have written down on the back of a misplaced envelope. Sweden has a commitment in law to reach net zero by 2045, and a recent momentous legal ruling in Germany has meant their target has been brought forward by five years to 2045.[4] The UK, Canada and New Zealand have legislation for net zero by 2050. Scotland, too, has set a net-zero target of 2045, because if there's one thing Scotland likes to do, it's to get it right up the English. If petty rivalry solves climate change, then I am absolutely 100 per cent behind it.

The EU and South Korea have both proposed legislation for 2050. President Biden has re-joined the Paris Agreement, and the United States, too, aims for net-zero emissions by 2050. Biden has also announced the American Jobs Plan, which will spend $2 trillion on infrastructure, about half of that going on long-term climate jobs.[5] Perhaps most importantly of all, China has committed to peaking* its emissions before 2030, and being carbon neutral by 2060. However, recent analysis suggests China must close 600 coal power plants by 2030 to stand a chance of meeting this target.[6] China accounted for

* Yes, I did consider using Peking, but I am above that sort of shit.

85 per cent of the world's planned new coal power plants in 2020.[7] The fastest proposals for achieving net zero are by Finland for 2035 and Uruguay for 2030 (the latter will simply threaten that Luis Suarez will bite anyone who doesn't comply), though neither are yet in law.* Many more countries are putting targets in place – although India, currently the world's third-largest emitter, has yet to commit to a net-zero target. It is quite a touchy subject there. Hopefully once I've pointed out the ways in which climate change affects cricket, they'll soon be on board.

Now, it's all very good having a target for twenty or thirty years' time, but what are countries doing *now*? Politicians love setting long-term goals that sound great, but putting them into practice is another thing entirely. Just like I love saying I'm going to lose a stone by next year, just before opening another tube of Pringles (except the future of humanity does not rest on me not eating crisps). Meeting the long-term temperature goals of the Paris Agreement requires countries to set short-term goals in order to get there. These are called Nationally Determined Contributions (NDCs). Unsurprisingly, at present, these commitments fall short. A 2020 UN report estimated that ambition will have to be tripled to meet a target of 2°C, and increased fivefold to set us on the path towards 1.5°C.[8] In other words, everybody is still chowing down their crisps.

Helpfully, the UK's Climate Change Act has this exact 'short-term targets to meet long-term goal' framework embedded into it. Handy! Legally binding climate budgets are set by Parliament after advice from independent advisory body the

* Talking about fastest proposals – Ian has just found out that his ex-wife is now engaged to that 'sweaty lycra-clad tosspot'.

Climate Change Committee* which acts like an angel on the government's shoulder, whispering 'You really need to get your act together, hurry the fudge up'. The UK has subsequently increased its ambition to a 68 per cent reduction by 2030 compared to 1990 levels.[9] The sectoral targets announced in the government's ambitious ten-point plan – the phase-out of dirty cars by 2030, quadrupling offshore wind by 2030, and installing 600,000 heat pumps a year by 2028 – will go some way towards getting there. Yet again, though, stating targets is the easy bit. Saying, 'We're going to do this, this and this,' is fine – but *how* are you going to do it is the single most important question that needs answering. Putting the actual support and regulations in place to meet these goals is a pivotal task. In principle, the 10-point plan is mostly saying all the right things on green solutions. Sure, some of these policies, such as 'Jet Zero' air travel, are clearly only in there to make the numbers up to ten – and because we have a prime minister who likes puns. Yet, at the same time, there's not really joined-up thinking about what we need *less* of, e.g. greenlighting a new coal mine, building new roads, and allowing more North Sea oil and gas exploration. Other countries such as France and Spain have banned new oil and gas exploration and set dates to end production.[10] I'd like to see the UK and Scottish Governments, as the hosts of COP26, follow suit. For the UK, it appears that saying no is a lot harder than saying yes – a bit like dealing with an unruly child. There's no wiggle room or time left for this sort of shit. So, the jury is still out, in my personal opinion.

---

* Committee on Climate Change.

Adapting to change and putting money on the table

Many countries are also putting in place measures to start adapting to a changing planet. National Adaptation Plans are being put together by countries to help them prepare for the inevitable changes already in the pipeline, and to make themselves 'climate resilient' – which basically means thinking two steps ahead, so you don't get royally screwed. It's about making your country futureproof. For example, green infrastructure – things like permeable pavements, rain gardens and trees to soak up all the extra water in cities – and farmers changing crops to ones better suited for future conditions. The required measures will be different for different countries, but nowhere will escape being impacted by climate change. And it's not even limited to just our planet – even Mars won't escape, as it will eventually experience an infestation of irritating billionaires.

The Committee on Climate Change's 2019 Progress Report states: 'England is not prepared for the impacts of a 2°C increase in global and UK temperature, let alone more extreme levels of warming.' Which isn't a surprise, considering most of us don't think to buy an umbrella until we're already being rained on, and we live in a country where it rains A LOT. Much of the UK's infrastructure dates from the Victorian era, and the privatisation of utilities has meant a short-term focus on keeping costs down rather than planning ahead. All in, it's a perfect storm for not dealing with storms.

How fair is it to force countries or people to have to adapt to climate change? And how fair is it to ask them to bear the brunt of an energy transition which can impact their way of life? There are questions of justice that pervade the climate movement. For instance, in northern Norway, indigenous reindeer herders are fighting against a large wind farm proposal that they claim will interfere with migration paths.[11] Many of these same Arctic peoples are already being directly affected by a warming climate and melting ice, forcing them to adapt. The Sami people rightly say that they have done little to cause climate change, and yet they are affected by both climate change itself and the steps being taken to mitigate it. And then there is the 'loss and damage', a term which describes the situation where adaptation will no longer suffice to buffer people against the impacts of climate change, and territories and livelihoods are lost for good.These people and places need direct support and to be treated fairly given the historical context of responsibility for climate change. Personally, I think the Paris Agreement does not go far enough in this respect.

One often neglected aspect of the climate debate is finance. That's mostly for good reason, because most ordinary people find finance tedious, and the people who work in finance are always trying to make up for the fact they are hugely overpaid for essentially meaningless work by undertaking sadomasochistic attempts for charity to make themselves feel less dead inside. 'Can you sponsor me for Tough Mudder?' Erm. No. It's not my fault you wake up empty. And I can say that, as I worked in finance – something my parents wish was still the case.

Anyway, climate finance *is* extremely important. To simultaneously move away from fossil fuels and towards renewable and other low-carbon options, we need a massive shift in investment. This doesn't happen by magic. An estimated extra $130 billion of green investments, roughly the entire worth of Bill Gates[12], is required every year until 2030 just to meet the current commitments. That amount may need to double or triple to meet 1.5°C.[13] Now I say that, I feel like maybe I shouldn't have slagged off all the finance people, as we'll probably need them on side. Access to finance is especially key for developing countries, who not only have contributed the least to climate change, but also have a higher risk premium for investment projects, making it comparatively more expensive for them to mitigate.[14] Therefore, access to finance has to be a key part of the UK-hosted COP26 in Glasgow, which is scheduled for November 2021.

We built these cities

Action is also important in regions and cities. During the Trump years in the USA, when federal level was moving in the wrong direction entirely, with the mere mention of climate change being removed from official documents left, right and centre, a vacuum was left that had to be filled at other levels. It galvanised others to make deep commitments. States representing 55 per cent of the US population and 40 per cent of emissions, from Colorado to Connecticut, from Maine to Massachusetts, from Vermont to . . . you get the idea, formed the US Climate Alliance, which committed to defiantly upholding the Paris Agreement.[15] California, which has an economy larger than the UK's, has stepped up and is making all new cars zero carbon by 2035, with a commitment to make the entire economy carbon neutral by 2045.[16]

Cities are also incredibly important, especially as over half of people worldwide live in urban areas. To be low-carbon, it helps if cities are compact. For instance, Seoul has a larger population than London but is a quarter of the size.[17] Alternatively, we could shrink ourselves, like in the Matt Damon film *Downsizing*, a depressing remake of *Honey, I Shrunk the Kids*, but about overpopulation and global warming. Cities also need clean mass public transport connections, like the world's first commuter cable car in Medellin, or the world's first fully electric bus fleet, in Shenzhen.

Copenhagen aims to be the world's first carbon-neutral city by 2025.[18] How is it doing that? Well, with a shit-ton of bikes, and by keeping the public on board by designing the plans with them, making the changes relevant to their everyday lives, and providing them with a positive narrative that everyone can believe in – plus, they built an artificial ski slope on top of a waste incinerator. Hell yeah! Oslo is moving towards having no emissions from its public transport by 2028.[19] Norway? Yes way.

Glasgow and Edinburgh are fighting it out to become the UK's first net-zero city, with the latter aiming for 2030.[20] It's going to be the biggest east vs west coast rivalry since gangster rap, with silent drive-bys in electric cars – and I am pretty sure it was Edinburgh City Council that burned down Glasgow School of Art. A few years ago, I met Manchester's Mayor Andy Burnham at the time when the city was looking into how to play its role in mitigating climate change, and was extremely impressed by the commitment to making sustainability part of decision-making. After considerable consultation, Manchester eventually settled on going net zero by 2038.[21]

Cities can step up to the plate when countries are doing little.

While Australia has been a bloody bludger on climate change, mostly telling the science to rack off, Sydney is doing 'heaps of hard yakka' by aiming to hit net zero by 2035.[22]

Business time

Companies also have a huge role to play, and some of them are showing signs of progress. Of the world's 2,000 largest public companies, just over a fifth have net-zero targets in place.[23] IKEA aims to be 'carbon positive' by 2030, which sounds good, but also semi-confusing given others are going carbon neutral and carbon negative. IKEA is offering customers vouchers for returning old furniture in order to promote more sustainable consumption and says it will make all its products from recycled or regenerative materials by 2030.[24]

In the UK, supermarkets are sharing targets for reductions too, with Tesco aiming for net zero by 2035 and Sainsbury's by 2040. Where will I do my net-zero big shop? Nespresso claims it'll be net zero by 2022: good luck with that. At present, its website says Nespresso will plant trees in regions where its coffee comes from. I imagine in some way George Clooney is involved.[25] Interestingly, a number of airlines, including British Airways, Cathay Pacific and Qatar Airways, have committed to going net zero by 2050. I have no idea how they think they are going to do that. Recycled paper airplanes, perhaps?

While these kinds of goals are crucial to set us in the right direction, we need to be extremely wary of how much '*net*' there actually is in all these claims of net zero. Always ask, how much is from your own emissions reductions, and how much is 'other stuff'? It is too easy to pretend to be net zero by simply paying for lots of offsets. Are offsets really the only way they can reduce emissions? If not, this is the equivalent of the net-zero drinker having a pint of Guinness but simultaneously paying someone else the cost of a pint of Guinness for them to not drink a pint of Guinness. You're still drinking the Guinness, mate! It suits the rich – and if everyone does it, then it doesn't work. If every company under the sun is offsetting their emissions through tree-planting schemes, I'm not sure there will be much room left on the planet for all of us to live. I for one cannot wait to live in a forest because my house has been bought to help a donut shop achieve climate neutrality. It'll become like that Talking Heads song: 'This was a Pizza Hut, now it's all covered in daisies'.

Call the COPS

The UK will be hosting the 2021 UN climate conference (COP26) in my hometown of Glasgow.* It is the most important COP meeting since COP21 in Paris in 2015, as it aims to help countries ramp up their short-term ambitions to 2030 by making stricter commitments – which, as I've mentioned, is bloody crucial. In an optimistic case where all long-term

---

* OK, technically not my hometown. As I said earlier, I'm actually from Paisley, but most people don't know where that is, so I just say Glasgow. I mean, come on: Glasgow Airport is *in* Paisley, FFS. Paisley is the last stop before Glasgow on the train. The phrase 'getting off at Paisley' was often used to describe a certain *coitus interruptus* contraceptive method of sexual intercourse.

country targets are met, i.e. EU and USA achieving net zero by 2050 and China by 2060, we could get to about 2.1°C. However, based on current short-term pledges, the estimate is more like 2.4°C.[26] The result of this COP will be called the Glasgow Agreement. I have drafted my own version to save some time, and hopefully the heads of government can simply sign this statement:

> *We, yon parties of the United Nations Framework Convention on Climate Change, are tellin yeez that the planit canny keep gettin mad wi aw this clatty carbin kicking aboot up there n that. We are absolutely buzzin tae agree that climate change can get tae, an promise tae gee renewables an that absolute laldy right noo. Belter. Yaldy.*
>
> *p.s. yer da sells avon.*

I imagine this will be signed by Billy Connolly and Sharleen Spiteri from Texas on a beermat in the back of the Horseshoe Bar, and afterwards all the heads of state will wander down to the Blue Lagoon chippy for a haggis supper and get into fights in the taxi queue outside central station. The fact that Glasgow may be the place that the world finally turns a corner fills me with a great sense of pride. And I hope more than anything to show ex-UN secretary general and all-round legend Ban Ki Moon around the Barras, and then take him to a St Mirren match.

In light of the important role Scotland is playing this year, I want to share a poem about climate change.

A poyum aboot the planit

A plonk masel doon on ma seat,
An am turnin oan the telly,
A KitKat for an evening treat,
Tae satisfy ma belly,

Up next noo on the BBC
Is David Attenbrother,
He's showin us aw a lovely tree
An a monkey wi its mother

But the monkees hame is under threat,
Coz the trees are comin doon,
And the forest isnae quite as wet,
So their food is turnin broon,

Aye the poor wee thing is so confused
As its mother isny well
An noo the forest's up in flames,
It looks like Dante's Hell

We've changed the wurld's thermostat,
By pittin oan the heatin.
We need tae find a way to help,
It's no good sittin greetin,

It startit off oh so weel,
Wi the Industrial Revolution,
But noo a monkee's maw is eel
Coz a aw the bad pollution,

The documentary's near o'er noo
Old Dave is on the screen,
'There isny much time left' he says,
'We're needing to go green.'

I channel hop oer tae ITV
But ma heid is still a'reelin,
Thinkin how does this relate tae me,
An what it is am feelin?

I'm hopin noo for sumthin light
But its halfway through the ads
A burger deal, a short-haul flight,
And SUVs for dads

An soon back oan tae BBC,
In time fur *News at Ten*,
There's floods in rural Lancashire,
We're back tae this again

Protester's scream, an cry, an shout
That the current system's brakin,
That naebday is doin nowt,
An the political will is lackin

Huw says: 'Here's Sophie with the weather'
'Thanks now Huw, it's aw gone mad'
'There's lots of sun, then rain and floods'
'It's looking pretty bad'

For an evenin then I've had enough
An soon am headin bedwards,
Still thinkin aboot this climate stuff,
An how a luv Huw Edwards.

24

What's going on in our heads?

Thinking about climate change is exhausting. It seems like a never-ending barrage of terrible news stories about massive world-changing scientific findings that inexplicably get worse and worse with every passing day. It's hotter, there's another fire, we only have so many years left. Jesus. How are we, as teeny, individuals, meant to cope with being bombarded by bad news on a constant ongoing basis?

And that's just the impacts. There's also the effort needed to *stop* all these atrocious events to consider. What do we do about it? How do we solve it? How can I help? It all seems so massive and brings up conflicting emotions and hard choices. One minute, you are feeling guilty about your own actions, the next it occurs to you that those individual actions don't even matter, and that you have no control whatsoever. It feels overwhelming. It's too big a problem, with too many messages and questions, and no obvious answer.

So, you close the browser tab with the depressing climate story and open NowTV to continue watching *Game of Thrones*.

Then your partner, who works on climate change, pipes up and says, 'Actually, this show is an allegory for climate change – you know, the whole "winter is coming" thing, that's actually blah blah blah . . .' and you zone them out completely and tell them that maybe they should go back to writing their book instead of ruining the only time of day you have to yourself. (This last bit isn't about anyone in particular.)

It's not totally our fault. We aren't really equipped to deal with climate change.

The Polar Bear Problem

Your brain isn't designed to grasp the issue of global warming. It's not because you, specifically, are a piece of shit. Various aspects of how we have evolved make it a tricky problem to say the least. For thousands of years, we lived in small tribes, where our brains were wired to respond to immediate dangers, like a fire or a lion or a lion on fire. Nowadays, the threats we respond to tend to be things like trying to avoid a sponsorship form at work.*

A changing climate, however, takes decades or even centuries. It happens so slowly we barely notice it. Like how Holly Willoughby is now omnipresent – when did that happen? As a danger, climate change feels distant from our day-to-day worries about whether we remembered to put the recycling bins out. For those less well-off, it is even further from the main concerns of putting food on the table or keeping our family

* I mean, its 5km – I literally walked that this morning. You're getting absolutely nothing from me, Phil.

warm. You can't stay constantly on edge about climate change twenty-four-seven, because we are not built to do that. That's no way to live a life. I mean, some people do – you know the type – but they would be like that anyway. For the rest of us, it slips from our minds, and so doesn't feel that dangerous. If it was really that bad, it'd be on the TV all the time and I wouldn't just be watching another episode of *Pointless*.

We are tribal, so we are interested in those closest to us, and we often mistake climate change as an issue only for people far away. It's not about us. It's about them. For instance, in the UK, 59 per cent of people think climate change is a very serious threat for people in less developed countries, but only 26 per cent think it seriously threatens them and their families.[1,*] The impacts we hear about are often happening far away, in the Arctic or in the tropics. They are not happening in Waitrose in the town centre.[†]

Even worse, when environmentalists try to draw your attention to the issue, their default is to show you a photograph of a starving polar bear on a sad iceberg. The quintessential image that everyone associates with climate change. It was on the cover of *Time* magazine in 2006 with the headline: 'Be worried. Be very worried.' And you say, 'My, my, that's terrible,' while your subconscious goes: '*Who gives a stinking crap, I've never even been to the Arctic. I don't even know a polar bear.*'[‡]

'An issue that already suffers from a lack of proximity has chosen as an icon an animal that could not be more distant

* And 43 per cent think it is very serious for the UK as a whole.

† They're not even happening in the big Sainsbury's.

‡ I do. My mate, Big Gary the Polar Bear, is one but he's shacked up with a grizzly now.

from people's real life,' writes George Marshall in his book *Why Our Brains Are Wired To Ignore Climate Change*.[2] Indeed, showing pictures of starving polar bears is fine if your audience are hardcore environmentalists,* but it's not necessarily going to get the message you want across to the average person† in the way you'd hoped.

Why climate change is like my dad

A major part of the problem is that we've evolved to care about things that feel close to us, whether it's the people we love or the dangers we face. However, like my father, climate change seems *distant* in many ways.

For one thing, the *way* climate change is talked about can make it seem distant from day-to-day life. It is boring and academic; it is for scientists. It also appears far off in terms of time. We're largely concerned about whether or not it'll be raining later today, while scientists are discussing what the global temperature is going to be in the year 2100. Nobody cares about the future. (Except my dad, who worked all his life in pensions. He loves pensions.‡)

We find it hard to imagine things we haven't experienced before. Having a baby never quite felt like it was real until it

---

* Or my mate Big Gary.

† Or seals.

‡ A joke of mine – 'I wonder if the inventor of the shoehorn ever tries to bring it up in conversation?' – was once retweeted by Martin Lewis, you know, the Money Saving Expert guy, and my dad thinks that's genuinely the single most impressive thing I've ever done.

was. As Marshall excellently puts it: 'We make it just current enough to accept that we need to do something about it, but put it just too far in the future to require immediate action.'[3] The problem is, we can't wait until everybody is experiencing the worst impacts before we do anything about it.

It's also distant in the sense that greenhouse gases are, like the hooded man with an axe standing behind your bedroom curtain right now, not visible. If we could see these gases, and they were, I don't know, bright pink, it would seem more immediate. And probably kind of cute. OK, what if they were grey. No, wait, that would just look like a grey day. I don't know, you pick the colour.*

What I'm trying to say is us *Homo sapiens* are very much of the 'out of sight, out of mind' tradition. Plastic waste is more visible to us, and therefore easier to remember on a daily basis. You see a discarded Coke bottle and think, 'That's harming the planet,' whereas when you see a car, you don't think 'That's harming the planet,' you think, 'That's a perfectly normal everyday car.' Or, if you're me, you think, 'That reminds me, I'd better pick up the new copy of *What Stairs?* magazine.'

Defence against the green arts. Facts are a blunt knife. Watermelons

A major problem has been the way that climate change has, at times, been communicated, both by scientists and the media. Coverage has focused on presenting more and more hard facts to persuade the public and governments to take action. If it's not

* What about red? Red sky at night, global warming!

working, then the plan has been to double down and present harder facts. As if people will eventually get it. They say, 'Sea levels are rising EVEN FASTER than we previously said, you idiots – it was three millimetres last year.' And we go, 'That's the size of an ant, so who cares?'.

But more information about how bad things could get is not necessarily the answer. People don't need to be *au fait* with the ins and outs of the science to know that it's happening and that something needs to be done about it. As Dr Kris De Meyer, a neuroscientist at King's College London, says, 'There are lots of situations in life where we don't have all the knowledge, and we still deal with it perfectly.' We don't need to know all the different ways that we could die in a car accident in order for us to keep ourselves safe, he says. 'It is the same with climate change. We don't need to know everything about bad things that could happen to us in order to deal with it.'

There are also other factors at play. When facts come up against tough obstacles, like the values we hold or the lifestyles we enjoy, then often it's the facts that lose out. It's not like we're out-and-out evil and actively want animals to die; we just like barbecues with our friends.

Facts are therefore often ineffective. They are like those big soft-play pugil sticks they battled with on the TV show *Gladiators*. They are never going to get through people's armour. Whereas using stories and emotions to make your point is the equivalent of taking off the soft parts and replacing them with sharp blades. Now Wolf is bleeding from his eye – but he's listening.

Our brains are clever at shielding us from uncomfortable facts and truths. They come up with various ways to cope with the cognitive dissonance of holding opposing positions. 'I always drive to the local shops to pick up the paper, even though I know this is damaging the planet.' This can be a difficult one to sit with in our heads; it makes us feel bad. To solve this internally, we explain the fact away or we ignore it.

This dissonance leads to a variety of responses where we are unwilling to accept the reality of the situation. Norwegian psychologist Professor Per Espen Stoknes suggests in his book, *What We Think About When We Try Not To Think About Global Warming*, that, 'maybe it would be better if we stopped speaking of climate denial altogether, and rather spoke of climate resistance.'[4] We say things to make ourselves feel comfortable and to justify ourselves. Like, 'I do other green stuff – I cycle to work once a week, so driving to the shops is fine,' or, 'The science isn't totally settled,' or, 'It is actually China that is the problem.'

At the global level, we have to deal with the conflicting knowledge that the planet is in danger, yet nothing seems to be being done about it. That worry can make us freeze, meaning inaction breeds inaction. Once this happens, various forms of denial are given the space and oxygen to take root. We will come to see how these feelings are fed, watered and grown by vested interest groups in chapter 26.

How did this purely scientific issue become tainted with social meaning about cultural identity? The answer comes down to politics. The solutions to climate change require what appears to be significant government intervention. Markets don't, by themselves, internalise the harm done to society by climate pollution, so we are required to tax or regulate things we want less of, i.e. carbon. In economics, this is simply correcting a negative externality – when someone's actions, that benefits them, harms others. But because it requires significant government intervention, many on the political right assume it's a left-wing scam to introduce the things they really want, like re-nationalising All Bar One.

In America, environmentalists are often referred to by libertarian sceptics as 'watermelons' – green on the outside, red on the inside. The term paints environmentalists as secret communists, using 'green' concerns as a smokescreen to push other agendas. The truth is that it doesn't really matter if climate action is sometimes used as a smokescreen or not; we still need to solve the problem.

Being aware of these framings can help avoid communication pitfalls, and shows why new voices, e.g. conservative and religious spokespeople, are important to convince those that do not believe climate change is a critical issue. We are more likely to listen to those that are like us and people we trust. You really do not need to be an environmentalist to care about climate change. In fact, it has perhaps been to the detriment of wider acceptance that climate change has been considered an '*environmental*' problem and not an '*everything*' problem.

Most who oppose climate science don't do so because they are ignorant. They do it to protect themselves from a perceived threat to their way of life, and to express who they are to those around them. If rejecting government intervention is part of your cultural identity, then it may feel like more of a risk to become a social outcast by accepting climate change than to be wrong about climate change. If you love driving fast cars or mining bitcoin, and have built your entire personality or career around the fact you are an absolute nerd about these frankly bizarre activities, they will win out. The threat of your mates saying, 'You've changed, man,' feels more immediate than a heatwave. Here, confirmation bias can kick in and those in such groups seek out information that corroborates their pre-existing beliefs, while finding ways to dismiss the rest. Also, it is simply embarrassing to be wrong. Unfortunately, social media nowadays makes it much more public.

I am not Jesus Christ

Messaging around climate action has typically been pretty negative. It says to *stop* doing the things you like – eating rib-eye steaks, flying to Ibiza, stealing cars – to prevent the apocalypse. At its most extreme, it is also about shaming. In a way, it sounds like religion: we have to quit our sinful pleasures now in order to achieve some far-off better future. And if we don't, then we'll end up somewhere much hotter than we would have liked. Also, I don't even know whose side God would be on when it comes to climate change, as we all know The Big Guy™ loves a whopping great flood.

When we do take small climate actions, like reducing our meat consumption, it can sometimes make us feel more inclined to take other green steps, but it can also make us more likely to feel that we've already 'done our bit'. I know I have been guilty of this sort of thing in the past. A wonderful example of this thinking was on the last day of the Edinburgh Fringe in 2018. After my show, there was a queue of about fifteen audience members outside, who all wanted to ask questions, chat or say hello. I eventually got to the last person on the last day of the last show. And the nice young woman, who had waited so patiently for about fifteen minutes, said: 'If I do all the things you talked about – stop flying, drive less, go vegan – then how many dogs can I have?'

Another religion-like aspect is in the oversimplified assumption that everyone is either a convert or a denier. I've had reviews for shows saying that I'm 'preaching to the converted'. Now, I'm not a preacher as such. I see myself more as one of you, a normal man. It just happens that those people come and gather round my feet to learn from my teachings. OK, I am not saying I'm the son of God, but maybe I am a bit like Jesus. I'm in my thirties and trying to save the world, I love Easter eggs, I've got a beard, I like red wine. I only have twelve friends, one of whom I don't trust. Plus, my dad is very controlling and does nothing on a Sunday.

It's easy for people to imagine that the whole world is split in two: converts and deniers. Those are the people you hear the most from. But in reality, the general public is considerably more nuanced than that. The Yale Program on Climate Communication has been tracking public attitudes for a decade now. It shows that there are actually six distinct groups when it comes to climate change, namely Alarmed (26%), Concerned (29%), Cautious (19%), Disengaged (6%), Doubtful (12%) and Dismissive (8%).[5] Coincidentally, all these are also terms my dad has used about my career over the last year.

Even combined, those in the Doubtful and Dismissive groups are only around a fifth of the population. And this is America, the home of climate denial. The problem isn't denial. The problem is that over half of the public (those in Concerned, Cautious and Disengaged groups) have an attitude towards climate change similar to hearing there's a new U2 album coming out. Utter ambivalence. We need to focus much more on engaging these middle groups. We need to move their attitude to 'Alarmed' – as if they've heard the new U2 album will be downloaded directly on to their iPhone.*

Many of these people are more concerned about the economy or national security, both things that top lists before elections. We need to understand why, because these issues are not necessarily closer to our lives than climate change. What can we personally do about taxes or terrorists? How many of you are actively trying to become the Governor of the Bank of England or the Head of MI5? What are you doing for national security (apart from putting a tiny Head & Shoulders bottle in a little plastic bag at the airport)?

The truth is we can do quite a lot about climate change compared to these other worries. I often ask audience members where they feel they sit on the spectrum. Concerned is the most common answer. I asked one young man why he wasn't Alarmed. He said it sounded exhausting.

Smells Like Teen Spirit

How do we make it easier to break down the barriers we have on climate? Well, some aspects of what I've described above have already changed in the last few years. In 2018, a number of events coalesced to massively increase public appetite for

* I would consider my mate Ian to be in the 'Cautious' group, as often he seems worried about climate change but then quickly changes the subject, mostly to how he's added cheese to his recipe, so it's now toast under beans under cheese. He recommends Co-op's own-brand mild cheddar. I recommended he tried a vegan cheese. He looked at me with disdain and simply walked away. It was the first time since the divorce I noticed a sparkle in his eye.

climate action. If 2020 was the year of coronavirus, then the prior year was dominated by a warming planet. Alongside VSCO Girls and Q-Anon, climate change seemed to go mainstream in 2019. People were seeing temperatures that made them uncomfortable, which tallied with what scientists had been saying would happen for years. Sir David Attenborough released beautiful documentaries that captured the public's mood across all walks of life. There were protesters blocking bridges and children marching on the streets demanding climate action to protect their future. Talking about climate change had finally become acceptable in society. At the Edinburgh Fringe in 2019, almost every single comedian's show had at least had a passing reference to climate change, which was very much not the case in previous years.*

Young people have joined the conversation in a big way. Teenagers today have never known a time without concern about climate change, and have come of age during the warmest decade in human history. De Meyer compares this to how many teenagers in the 1980s had anxiety about the very real threat of nuclear holocaust. As a teenager, you are at an important developmental stage, he says. 'Our brains have critical development periods, where things become embedded much more deeply, than [they would] if we were to be exposed to the same message when that critical development window has closed.' That is to say, for many teenagers who have grown up with an understanding of climate change it is not some far away thing, it's here and now and will affect them.

De Meyer tells me the following story of one of his students. While teaching in 2018, he presented studies to his class which

* Get your own gimmick.

indicated that people reported a higher concern about climate change after watching the 2005 film *The Day After Tomorrow*, but they did very little to act on this concern when followed up days or months later. 'After that class, one of the young students came to me and said, "You know, I watched that film when I was twelve. And it's the main reason why I'm here in this class." And I thought, OK, these young people who saw this when they were young, it had a very, very different impression on them.'

The story illustrates that our brains play an important role in all this. Hearing a message when you are fifteen, when you're developing your capabilities for processing moral and social reasoning, is different than hearing it when you're thirty or sixty. They have grown up understanding what climate means, the threat it poses to their whole lives, and they are rightly concerned.

Bringing it home

We need to focus on telling stories about how climate change is affecting people like us, near us, here and now. And even more importantly, we need to be telling positive stories about people, like us, who are tackling climate change. To provide examples of what can be done in our communities. To thaw the freeze on climate action, and inspire more people to become Alarmed. But not alarmed in a debilitating, doom-laden manner; alarmed in a motivated and engaged way.

We must also tell more stories of those around the world who are already suffering due to climate change. These stories need to be made more human and relatable to the audience in question.

And we need more voices from across the whole spectrum of society. As Professor Katharine Hayhoe, a climate scientist and Christian who has had success talking to people of faith about climate change, says, 'Nearly everyone already has the values that they need to care about climate change: they just haven't connected the dots between them.'[6]

We need to make solving climate change about taking positive actions rather than giving stuff up. It should be about taking new train journeys and local holidays, trying exciting recipes and making a positive contribution. You're not giving up a petrol car, you're getting a futuristic electric cybermobile, like Buck Rogers or *The Jetsons*, or some more modern science fiction reference. We need to discuss the optimistic version of the future we want. As others have said, Martin Luther King inspired people by talking about having a dream, not a nightmare. While hearing about other people's dreams is usually my idea of a nightmare, it was actually a pretty good dream. Providing space for meaningful action on climate change is essential in achieving the future we want.

If all else fails, take heed that I successfully managed to make my dad care about climate change when I explained how it could affect his pension.

'I never thought about that, son,' he said. 'Makes you think.'

25

Six months

I won't be afraid – I trust the process

I have sat down to write this entry late at night while simultaneously watching Oscar on the baby monitor. It feels more tense than watching Andy Murray in a Wimbledon final. Every single small movement is getting huge reactions from me. I'm like one of the line judges, hunched over perfectly still and watching intently. Barely moving and typing quietly in case I disturb him, despite the fact that he's at the opposite end of the house and there are three doors between us. At any moment, he may wake up crying. And, like a ball-boy, I will have to go pick him up . . . Oh, wait a minute. Shit!!!

OK, I'm back.

Yesterday, he made his first joke. Over the past months, he has been laughing lots and enjoying me stupidly dancing jigs, and his mummy blowing raspberries on his tummy. He once had a laughing fit for ten minutes after he watched a balloon being blown up and let down. Which, to be fair, does make a funny sound. But today he was the one intentionally making us laugh. He has worked out how to do 'peekaboo'. An absolute classic of the genre. He pulls up his bib over his face and then back down again, and gives a cheeky smile. And he'll do it again and again, with a level of commitment and repetition that even Stewart Lee would be proud of. He even pulled my finger, but farted himself – already playing with the form.

You always know you will love your child when they arrive. But it is an abstract love until that specific child is actually there, in your home, day in, day out. I think that has been the greatest surprise: that I didn't realise he would have his own personality right from the start, and more and more so with every single passing day. How quickly they become *them*. Before having a baby, I remember going to the christening of a six-month-old and thinking, 'Aww, they're so cute – but basically, they're just a baby, a void. They're not a person yet.' And now I realise how completely wrong I was. If you spend every day with a baby, you get to know and understand them deeply. Now I love *him*. All his little traits. How he scrunches his wee nose when he's excited. His one little dimple. The fact he defecates at exactly 7am every single day. I cannot imagine the world without *him*.

Yet, sometimes, I feel torn between baby Oscar and future Oscar. That is to say, I feel like I am dithering between being a dad and saving the planet. Daddy Matt Winning and climate researcher Matthew Winning are at odds, fighting between paying attention to him today and protecting his tomorrow. And failing at both. I have been working from home since he's been born. And while the benefit of seeing him so much undoubtedly outweighs the negatives, there are negatives. I do feel constant guilt that I am either neglecting my parenting or my work. Which version of me should be given priority at what time? Captain America never had to stop fighting bad guys because he had to do bath time.

Finding this balance has been a challenge. Hearing Oscar crying through the wall that his nappy has leaked everywhere and simply having to turn up the volume on my headphones is tough going. And I feel guilty that my wife has essentially put her own career on hold just now, while I sit in another room twelve hours a day, when she is clearly far more talented than I am. I wish I could give her more time to get on with her own work. Is writing this stupid book about climate change more important than being a good dad, or husband, or than my wife's career? Yet often when Oscar gets upset or is laughing hysterically, I get distracted, and I'll go in and see what's happening, either because I'm worried about them, or simply suffering from FOMO. I'll take him into the front room where I work, to distract him and give my wife a tiny break. We will look out the window and play a game I have invented to calm him down called '*Is That Car Too Big?*'. I ask him if the next car will be unnecessarily large. He gets to choose by touching my right hand – 'yes' – or my left hand – 'no'. Hours of fun.

So much of the conversation on climate action is about what sort of world we will be leaving behind for the next generation. And sure, the earth should be passed on in a state that, if sold second-hand on Amazon, would be considered '*Good As New*'. But it's also about what we provide *now*, today, for young people, for their whole lives. Fairness across generations – old and young. Fairness across borders – poor and rich. Human rights and climate justice.

The darker thoughts often seem to creep in when I am astoundingly tired. Sometimes I feel like I am utterly letting future Oscar down. I now have this duty of care that I didn't have before. A more real attachment to the climate crisis and a new motivation to ensure the world is moving in the right direction to a safe place.

I look at great things others are doing to defend the young from environmental harm. The UK Youth Climate Coalition is giving a voice to young people. In the USA, there is the Sunrise Movement, who are doing similar work, and also a non-profit law firm called Our Children's Trust that is dedicated specifically to taking legal action to protect children's and future generations' rights to a safe climate. They have ongoing actions in seven states and against the US Government. Greta Thunberg is still striking every Friday. The youth are speaking up and a movement has been born. It will eventually change the world. But eventually is not soon enough. They cannot wait. And they cannot do it alone. They need all of us to fight for them and with them. Young people are not yet the decision-makers. And time passes so quickly. The next decade will define the journey of our families, and of the human race, for millennia. Yet, here I am, singing 'The Wheels on the Bus' over and over for most of the day. I mean, at least it is about public transport.

I worry that someone who has slept more than I have would be more competent at saving the world. And I think it's the lack of sleep making me doubt myself. Captain America did such stellar work and always seemed well rested. I mean, I guess he did sleep for half a century. I can barely look after my family, let alone take on any responsibility for the end of the world as we know it. I'm not a superhero. They don't exist. I'm just one person. Who am I? At times, working on climate change can make you feel mad anyway, like you're the person with their finger stuck in the dyke, or that you're Cassandra.* I guess this is just a new variant of that madness. Also, the sleep deprivation is giving me delusions of grandeur and making me think I'm an Avenger.

* Rodney's wife from *Only Fools and Horses*.

26

Who's delaying action?

Now we have reached the *Scooby Doo* bit of the book, where you find out who was really behind the mask all along. And, unsurprisingly, as with *Scooby Doo*, it is almost always an old white man who owns a business. And he would have got away with it, too, if it wasn't for those pesky kids going on climate strikes.

A mix of vested interests, ideology, media bias, dirty tactics, psychology and flat-out bullshit have played a role in taking a fairly straightforward settled scientific issue and turning it into one of the most controversial topics in human history. But as you'll see, denial is simply part of an overall effort with the end goal of *delay*, using the tactic of *doubt*.

The Past

They Knew

Action on climate is a bit of a Catch-22. The public tend to say it is up to governments to regulate and for businesses to act.

Governments say they'll regulate when it won't hurt businesses. And businesses reliant on fossil fuels will say that it is up to the governments to force changes and they'll do it when there's a public appetite for it. And on we go round in endless circles.

Now, for business, this is perhaps a selfish but fair enough position to take. Except when your business also spends millions lobbying to stop government interventions and undertakes disinformation campaigns to mislead the public into believing the issue is highly uncertain when you know it is, in fact, certain. Then, perhaps, you are the one actively perpetuating that destructive circle.

The fossil fuel industry knew about the potential impacts of global warming long before you or I had even heard of it:

> [C]arbon dioxide is being added to the earth's atmosphere by the burning of coal, oil, and natural gas at such a rate that by the year 2000, the heat balance will be so modified as possibly to cause marked changes in climate beyond local or even national efforts.[1]

The words are from a 1965 speech by Frank Ikard, the then-President of the American Petroleum Institute (API) – the trade association for the oil and gas industry. Many of these companies themselves undertook a significant amount of research on the topic.[2] In 1982, Exxon estimated that atmospheric CO_2 could double by 2060, leading to a rise of over 2°C, and Shell's research in 1988 suggested it could happen much earlier.[3] I imagine their response was to lock these research results in a safely stowed cabinet and burn that cabinet to the ground using fossil fuels.

Then, in the late 1980s, at exactly the same time as global

warming started becoming apparent to the general public, some of these massive, powerful companies decided to shift their focus. The new emphasis would be on delaying action, saving themselves using the same tactics established by other dirty industries decades before.

How to go about this? Firstly, there is power in numbers and speaking with one voice, so you set up a nice-sounding organisation. And so, in 1989, the gentle-sounding Global Climate Coalition (GCC) was established.[4] Coincidentally (strokes chin) this was just as a movement on addressing global warming was forming, i.e. the year after the IPCC was set up and the headline 'Global Warming has Begun, Expert Tells Senate' appeared on the front page of the *New York Times*.[5]

Now, from GCC's name, you'd think they were a *Scooby*-gang of right-on activist do-gooders who really care about the planet, and not necessarily a lobby group whose members included the biggest oil, gas and automobile companies, along with every other company under the sun with a business reliant on fossil fuels, including the aforementioned API. Similarly, if you got a couple of people together and formed a group called the Citizens' Grasshoppers Alliance, it could provide a veneer to suggest that they know and care about grasshoppers, despite it later being discovered that you were actually a gang of rogue children with magnifying glasses burning insects to death.

The GCC group was influential in lobbying at UN climate meetings (no, I have no idea why they were invited either), often sending more delegates than some developing nations. It also financed commercials criticising a global climate agreement. BP and Shell left in the late nineties, and soon after it disbanded – the fossil fuel equivalent of Geri Halliwell leaving the Spice

Girls, only the type of 'power' the Spice Girls were pushing didn't cause asthma in children (that we know of). However, before its demise, the GCC was cited as crucial in persuading George Bush to abandon the UN Kyoto Protocol.[6] Job done.

Casting Doubt

The next step in the bucket list for delaying action is emphasising evidence that casts doubt.

'We shouldn't do anything about it until we are sure,' was the repeated mantra. Except we *are* sure – scientists, that is – but if the public believed we weren't, then the appetite to 'do something' would never arise. Gas companies gaslighting.

One example is a campaign by the Informed Citizens for the Environment, whose mission was to 'Reposition global warming as theory (not fact)'.[7] This shows the importance of language and public understanding of science, because technically, all science is a theory. Gravity is a theory. But to the public, talking about something as a 'theory' makes it seem less certain, even though it isn't.

Another example is an internal draft memo from the American Petroleum Institute in 1998 (yes, the same API that knew how bad it would get back in 1965), stating that:

> 'Victory will be achieved when [. . .]
>
> - average citizens "understand" (recognize) uncertainties in climate science;" and when "recognition of uncertainties becomes part of the "conventional wisdom".'[8]

This approach of cultivating doubt and uncertainty in the public's mind, to prevent strong climate action, mimicked and employed tactics used by the manufacturers of other harmful products, and even at times using the same contrarian scientists.

A number of the same scientists applied the approach of seeding doubt to many topics that would require regulation, including second-hand smoke, CFCs and the ozone, and acid rain.[9] There are examples from back in the 1930s where the asbestos industry used scientific uncertainty and supressed studies showing the harm asbestos could cause in an effort to protect their profits.[10] My grandfather died from asbestosis.

But oil companies cannot always cast doubt directly. If XYZ Big Oil Co. puts out a report that says maybe global warming isn't happening, the public wouldn't believe it. Instead, you have to quote someone else – someone that sounds believable and trustworthy. And this is where the fancy-sounding 'independent' free-market think tanks come in. Along with some plucky contrarian scientists who either take funding from the oil industry or are ideologically predisposed to them and loose with their scientific methods (or both). These think tanks provide helpful research, reports and public statements that muddy the water. They are all called something like the Something Something Institute, and conduct non-peer-reviewed publications that are then parroted in the media.

While a number of fossil companies began in the late nineties to distance themselves from outright climate denial, others did not. Greenpeace investigations found that ExxonMobil gave over $30 million to climate-denying organisations between 1998 and 2014.[11] And if you don't believe Greenpeace because you think that's the kind of thing Greenpeace *would* say, then even according to the UK's renowned historic scientific institution The Royal Society, ExxonMobil gave $2.9 million to thirty-nine organisations that 'misrepresented the science of climate change by outright denial of the evidence' in 2005 alone.[12]

The story of Dr Ben Santer is another good example.[*] Back in the innocent days of 1996, Santer found himself at the centre of a storm after working on the Intergovernmental Panel on Climate Change's Second Assessment Report. He was lead author of a chapter that had basically found the smoking gun of evidence showing that we were the ones causing climate change (i.e. the human fingerprint).[†] This was it. The guilty party had been found, namely fossil fuels burned by humans.

For those whose wallets would be hit if humans stopped burning fossil fuels, some doubt was required to throw at this finding. And as it just so happened, there were a few scientists who were experienced at pulling off exactly this sort of thing. Dr Frederick Seitz attacked both the IPCC, and Santer specifically, in a *Wall Street Journal* article. Now, Dr Seitz wasn't a climate scientist or anything, nor had he been involved in the report, but he had been a renowned physicist during World War Two and president of the National Academy of Sciences back in the 1960s. He appeared to be a man of science. However, after retirement from academia, Seitz had worked as the principal scientific advisor to the R.J. Reynolds Tobacco Company medical research programme, arguing tobacco wasn't harmful. In

* This story is told more extensively in the seminal book *Merchants of Doubt* and more recently in the podcast *How They Made Us Doubt Everything* on BBC Sounds.

† Remember the bit from chapter 4 where we knew it was coming from us because it was the *underside* of the atmosphere that was warming, not the outer layer, hence the heat must be coming from the earth? That!

1984, he co-founded the George C. Marshall Institute, a think tank which promoted climate-sceptic views.

In the *Wall Street Journal*, Seitz accused Santer of gross personal misconduct and stated: 'I have never witnessed a more disturbing corruption of the peer-review process than the events that led to this IPCC report.' In practice, the text required some standard procedural changes, that was all. It was an act of an incredibly corrupt pot calling an innocent kettle corrupt. But Seitz made out in a national newspaper that the IPCC had scammed the system. And, obviously, this was echoed elsewhere. The Global Climate Coalition – you remember them, basically ALL OF FOSSIL FUEL BUSINESS – sent a document to the press accusing Santer of manipulating the peer-review process, and of 'scientific cleansing' – echoing the ethnic cleansing that was happening at the same time in Bosnia.[13] All of these accusations were entirely unfounded, yet Santer spent months having to defend his integrity, and it led to dead rats on his doorstep and the break-up of his marriage.

Seitz, who was, as I mentioned, a former president of the National Academy of Sciences, would later help create a 1998 petition so misleading about climate change that the National Academy of Sciences itself had to publicly refute. The so-called Oregon Petition claimed that something like 30,000 scientists had signed up arguing against the reality of climate change. It turned out that most signatories were not climate scientists, others were duped into it, and many were even fake.[14] For instance, one of the signatories was listed as Geri Halliwell. Now, it probably wasn't her, as far as I'm aware she doesn't have a scientific background, but if one of the Spice Girls *was* a climate denier, then maybe it would be Ginger.[15]

False bias and fake scandals

And that brings us to the next step in the process, a naive and unsuspecting media that perpetuated non-existent uncertainties in order to appear unbiased, while unwittingly providing bias by doing so. Now, sure, a few media outlets already had certain ideological leanings and were happy to lean on them. But so many more spent decades providing what they believed was necessary balance on a story that didn't actually have two sides. There weren't the Sharks and the Jets – it turns out there were only Jets. The Sharks shouldn't have been given airtime.

The coverage played directly into the hands of those who sought delay. Typically, a climate scientist and a naysayer would be put head-to-head. Lord Nigel Lawson, the ex-Chancellor of the Exchequer and creator of the climate sceptical 55 Tufton Street think tank the Global Warming Policy Foundation,* appeared regularly on the BBC to discuss climate change despite having no professional expertise on climate change whatsoever.[16] This gave airtime to marginal views that often went unchallenged. He was improvising so much bullshit that instead of the *Today* programme, they really should have been putting him on *Just a Minute*.

The very existence of debate on TV and radio made it seem like there was one. It was basically the equivalent of having a COVID denier on every TV section about the pandemic.

Others made good use of print media. Exxon regularly ran adverts that looked like op-eds in the *New York Times* between 1989 and 2004. A 2017 study found over 80 per cent of their studies and internal documents acknowledged man-made cli-

---

* He is also the father of a famous chef whose first name suggests he either has no imagination or wanted a boy.

mate change is real, but only 12 per cent of their public adverts did so, with 81 per cent instead expressing doubt.[17]

Then, in the mid-2000s, we appeared to go through a sort of false awakening on climate change. The peak arrived with Al Gore's film *An Inconvenient Truth*, which hit screens in 2006. In response to the film, the Competitive Enterprise Institute ran TV adverts with the tagline 'Carbon dioxide: they call it pollution. we call it life.'[18] In 2007, Gore shared the Nobel Peace Prize with the IPCC. There were more media articles on the topic, and TV shows spoke about climate change regularly. There was massive momentum as we headed towards the end of the decade, with David Cameron running for prime minister under the banner of promising to be the greenest government ever.*

Things were getting too positive. So the 'forces of darkness'™ were called on once again to sow doubt.

In 2009, a few weeks before the UN Copenhagen Climate Summit, another scandal *just happened* to occur. Except it wasn't a scandal. It just had a good name – 'Climategate'. Essentially, thousands of emails of some climate scientists were hacked, shared online, and phrases were taken out of context to make it seem they were manipulating data sets in order to prove climate change was real. Which was complete nonsense, except it led to unfounded accusations of scientific impropriety being parroted around the world's media, as well as investigations into wrongdoing. All of this undermined credibility in climate science, despite no misconduct having taken place.

In particular, a sentence from an email from the Director of the Climate Research Unit at the University of East Anglia was given undue prominence: 'I've just completed Mike's Nature

* In hindsight, I presume by green he actually meant money.

trick of adding in the real temps to each series for the last 20 years (i.e. from 1981 onwards) and from 1961 for Keith's to hide the decline.'[19]

Climate sceptics suggested that the phrases 'trick' and 'hide the decline' meant there was a conspiracy at play. However, in practice, the former simply referred to a mathematical shortcut technique or 'trick' used in a previously published study. And the 'decline' mentioned was not about global temperatures, but with the fact that a tree ring dataset did not match reality, and therefore had to be reconciled.*

Basically, both accusations were boring non-events. Regardless, these phrases were repeated out of context in countries across the world, because to someone who knows nothing about the topic or the context, they sound nefarious. Media organisations across the entire spectrum, from right-wing to mainstream, ran stories suggesting there had been potential impropriety. Politicians and pundits called for the scientists to be criminally investigated, and some received emails threatening their families.[20] As I've said, all the investigations that were carried out found no wrongdoing on the part of the scientists.[21] To this day, nobody knows who was responsible for the hack, which was in fact the only crime committed – and it goes unsolved.

---

* For the sake of clarification. Tree growth is used as a proxy for temperature, mostly in the distant past when we didn't take readings in hundreds of places across the world. But in northern latitudes from 1960 onwards the growth declined and did not work as a proxy when compared to all the direct thermometer temperature measurements that scientists been taking during this time. As other factors beyond temperature – such as droughts, sulphur dioxide emissions, or air pollution – can also affect tree growth, scientists think one of these may be to blame. I think I fell asleep in the middle of this footnote it was so dull.

It was another case of 'attack the scientists to undermine the message'. At the Copenhagen summit, Saudi Arabia used the scandal to suggest that the science and scientists should not be trusted.[22]

A fascinating outcome in the wake of the Climategate non-debacle was the creation of the Berkeley Earth Surface Temperature Project, which was set up by some sceptical academics to independently calculate the temperature records from scratch. It was led by independent scientist Professor Richard Muller, who was himself highly sceptical about climate change, and among its funders was the Charles G. Koch Charitable Foundation. While there were already global records from NASA, NOAA and the Met Office, this project was intended to be even more comprehensive and go even further back in time. They would get to the bottom of it all.

Its findings came out in 2012 and what did Muller conclude?

'Last year, following an intensive research effort involving a dozen scientists, I concluded that global warming was real and that the prior estimates of the rate of warming were correct,' wrote Muller in the *New York Times*. 'I'm now going a step further: humans are almost entirely the cause.' He referred to himself as a 'converted sceptic'. That's right, yogurt fans – Muller changed his corner.

There were many other well-documented disinformation attacks on climate science. However, following the success of the Paris Agreement in 2015, the ground beneath us all began to shift once again. And when Donald Trump arrived in the White House in 2017, global political action elsewhere actually appeared to galvanise around the issue.

The present and the future

Same faces, new bullshit

The good news is that asking whether climate change is real is mostly a comical thing of the past, like shoulder pads and Spandex.

The bad news is that the push to delay action has evolved into something far craftier.

Now, the debate has turned to what we *do* about climate change. Today, in light of changed public perceptions, the 'climate inactivism' machine has had to develop into something quite different. It is no longer about denying the science (at least, that is, with the exception of Twitter accounts named @St3v3_4_Fr33d0m and mad cranks with blogs like Piers Corbyn, many of which have, unsurprisingly, recently turned their attention to COVID-denial). Companies that want to be taken seriously, including fossil fuel companies, can no longer deny reality. Public opinion has now shifted so far in line with people's experience that it is no longer good for business.

But that doesn't mean work behind the scenes has stopped. The largest five oil and gas companies are estimated to still be spending on average $200 million a year on climate lobbying.[23] As Damian Carrington from the *Guardian* puts it: 'So it's goodbye climate deniers, hello – and you'll pardon me for being blunt here – climate bullshitters.'[24]

Leo Hickman, editor of Carbon Brief, tells me: 'What we have seen is a move away from climate scepticism, to policy scepticism.' In some ways you could argue this is valid, he adds, because we want this to be a policy debate. But if one of the policy options is to not do anything, because it's not serious, that is a 'soft form of climate denial', he says. The old 'can't we just do nothing?' is a classic of the genre.

The debate is now all about the policies, the technologies; people asking 'is it too late?' and 'will it cost too much?' All of which is part of what Michael E. Mann, Professor of Atmospheric Science at Penn State, and creator of the hockey-stick graph, calls 'the new climate war'.[25] While that language could be considered rather inflammatory, when you consider the fact that he has been sent death threats and fake ricin in the post, you can kinda see where the guy is coming from.

The main goal, whether unconscious or not, is still delay. Doubt is still used, but now it is about the solutions. This is the stuff I mentioned earlier, about whether electric vehicles really have lower emissions, about the intermittency of renewables, about plant-based meat needing fertilisers and land on which to be grown. All of which are 'what aboutism' at its best. A prime example of this was the recent Michael Moore-produced film *Planet of the Humans* by Jeff Gibbs, which spent its entire running time spouting outdated or inaccurate or outdated *and* inaccurate clichés about green energy.[26] I presume they started making the film a long time ago, as many of the solar statistics provided were about a decade old.[27]

The film was released on YouTube likely because no film distributor or streaming platform would touch it. Because it was a stinky pile of lazy erroneous garbage. But people *see* this stuff. I was asked by a friend of a friend, who knows I work in this area, what I thought about it. He'd just watched it and didn't know what to think. Were solar panels bad? 'Surely it cannot all be that terrible,' he said. I told him that everyone who works in climate change reacted to *Planet of the Humans* in the same way as the rest of the world reacted to the film *Cats*. That nobody should be made to suffer it, and he should wipe it from his memory if at all possible. That *Planet of the Humans* couldn't have been any worse even if it had featured James Corden as a CGI cat.

Dr Kris De Meyer tells me that the same psychological mechanisms that led to people becoming entrenched in either accepting or not accepting the climate science, can lead to 'entrenched positions about what is the correct thing *to do* about climate change.' Even when solutions are being discussed in good faith, if we argue so much amongst ourselves, we risk splitting into unshakeable views about how to tackle climate change. This risks playing directly into the hands of those that wish to delay change. While we bicker, we continue to stall and divided we fall.

Another staple delay tactic is to say that it will simply cost too much to stop climate change. Now, it is absolutely correct to ask about the costs of climate action, discuss how these will be distributed across society, and talk about how and when to achieve reductions. This is literally my job, so, needless to say, people *are* thinking about it.

Overall, the impact on GDP of achieving net-zero emissions is likely to be negligible in the UK, and it could potentially have a positive impact on GDP.[28] Many of the costs are better thought of as investments anyway. However, if you only talk about the costs of mitigation without mentioning the costs of doing nothing, then you are only considering half the problem (and therefore living in a fantasy land). The UK Office for Budget Responsibility recently reported that 'the costs of failing to get climate change under control would be much larger than those of bringing emissions down to net zero'.[29] It added that the public debt involved in achieving net zero over the next thirty years will be less than the debt from COVID-19 was in two years. The important thing is to design policies to make sure that the transition is a fair one, so that those less well-off are not adversely impacted, and those whose jobs are affected are offered new opportunities.*

Doom and gloom

Another delay tactic is to get people to disengage by suggesting that it is too late to do anything. That society is about to crumble, and the only strategy left is to remove oneself and survive outside of society. This type of doomism has gained traction among some environmentalists. For instance, in a non-peer-reviewed, self-published paper about so-called 'Deep Adaptation' that has been downloaded over half a million times, Professor Jem Bendell from the University of Cumbria wrote: 'When I say starvation, destruction, migration, disease and war, I mean in your own life. With the power down, soon you wouldn't have water coming out of your tap. You will depend on your neighbours for food and some warmth. You will become malnourished. You won't know whether to stay or go. You will fear being violently killed before starving to death.'

* At the time of writing there has been a ramp-up of articles in the (mostly right-wing) UK press about the cost of net zero. It appears to be driven by some politicians feigning concern for the poorest being made to pay for net zero (despite rarely ever being concerned at other times). The impact of climate change will fall hardest on the poorest, but how they are affected by climate policies is actually a choice about how government implements and distributes the cost of policies. That is, they can make those with higher incomes take the burden – hence the likely real reason these politicians are opposed.

It's eerily similar to what my dad said would happen if I moved to London.

This type of doomsday-prepping, fatalist nonsense has, until now, normally been reserved for the more right-wing libertarian types. However, there is a growing trend of doomism within those concerned about climate breakdown. In a BBC radio interview in 2019, Extinction Rebellion founder Roger Hallam said, 'I am talking about the slaughter, death and starvation of six billion people this century.' An outcome that is breath-takingly made-up. It's not even original – it's just the plot of *Avengers: Endgame*. For me, a morbid obsession with outcomes of societal collapse will achieve nothing except supporting delay. And if I'm wrong, I'll let them eat me when the time comes.

Even the superbly written book *The Uninhabitable Earth* by David Wallace Wells, which essentially reads as a worst-case climate outcome, merely paints a picture of a world that, in all probability, will not happen. The truth is that action is not futile. As the great philosopher Natasha Bedingfield once said, 'Today is where your book begins, the rest is still unwritten.' Giving up to me seems like another form of denial. I constantly see people tweeting '*It's too late!*'. Well, it isn't. We are not driving off a cliff edge. The best analogy is that we are driving towards a brick wall, but it really matters how fast we hit it. There is still time to hit the brakes. But we need to hit them hard. That way, the impact will be minimal.

Humans will not go extinct. Neither is it likely that billions will die. I've had people say these two things to me, and seen the media quoting people saying them as if they are facts. It is true that some places and species cannot adapt, and will be lost forever. That's what happens when things change too quickly through a man-made process. Millions are already dying due to air pollution from fossil fuels. Suffering and loss for millions will certainly increase with a warming world. And yes, we must be aware of tipping points that could massively alter the state of the planet. But it doesn't need exaggeration or extra scaremongering. Perhaps some people are not concerned about the difference between millions and billions. Maybe they think it is better to motivate with fear than to be accurate. Perhaps it is just a lack of attention to detail, getting a bit carried away, or a tendency to fear the worse. I feel callous correcting people on that. Given the suffering already experienced by lots of the poorest around the world due to climate change, and will continue to go through to even greater lengths, I find Western societies pretending they will collapse rather insulting.

And what of fossil fuels?

Fossil companies are now positioning themselves as part of the solution. This has evolved into unrelenting greenwashing in the ultimate case of 'he who smelt it dealt it'. According to Sophie Marjanac, climate accountability lead at environmental law firm ClientEarth, 'Companies spend millions on reputational advertising to protect their social license to operate. That is, the ongoing public acceptance of their business practices.'[30] It turns out it's a whole lot easier to stick up a bunch of adverts than change your underlying business model. According to InfluenceMap, the five largest oil majors spent 42 per cent of their lobbying and branding budget on climate-related issues in 2018, even while company forecasts showed they only planned to use 3 per cent of their capital spending on low-carbon solutions in 2019.[31] Like a teenager who constantly brags about sex, despite the fact they've only done it once.

I remember going on the Eurostar to Brussels for work a couple of times during this period, and constantly being bombarded early in the morning with ExxonMobil adverts about making biofuels from algae. I was taken aback at the level of sheer arrogance required to spend money telling people you are looking into something which may or may not happen in about a decade. It's like me putting out adverts claiming I might be signing for Barcelona because I bought their strip and a new football from Sports Direct.

There was a similar BP advertising campaign, which I mentioned before, involving posters on the Tube about using banana skins for jet fuel. My poor wife patiently listened to a tirade from me for about forty minutes on the journey home one night about how this was 'an absolute crock of shit'. And not just because it ripped off the episode of *Bananaman* when he went into space. This exact Possibilities Everywhere campaign by BP was taken to task by ClientEarth lawyers who complained it misled the public by focusing on BP's low-carbon energy products. BP responded by quickly removing the campaign.[32]

Someone recently asked me what fossil companies *should* be doing, and what would it take to convince me that they are changing. These are good questions.

To stay roughly under a specific temperature target, there is only so much more carbon humanity can emit. We have already emitted 92 per cent of the carbon allowed if we want to stay below 1.5°C.[33] Without immediate rapid reduction from today we will surpass 1.5°C sometime around 2030–32.[34] Even with modest emissions reductions, a median estimate of when the 2°C threshold will be passed is by around 2052.

The carbon budget concept has helped bring into focus the vast reserves that fossil fuel companies have on their books that we cannot burn if we are to meet these global climate targets. Such 'unburnable carbon' is risky for oil and gas companies, some of whom are already having to write off assets as their values plunge.[35] You can see why it seems easier to pay a PR company to create some nice bullshit about powering a plane with plums.

Many companies are slowly making inroads into new territory such as renewables, clean energy provision and electric vehicles. Several companies in the fossil fuel industry have made commitments to reach 'net-zero' emissions by 2050, including BP, Total, Equinor and Shell. While it is both commendable and a pretty big deal that such companies are on board with the net-zero plan, they also sort of have to, otherwise their social licence to operate would probably be revoked.

There are obviously still issues. A 2050 'net-zero' target leaves significant room for interpretation for near-on three decades. Many companies' near-term targets for 2030 are based on the carbon intensity of their energy, not on actual total emissions. This quite opaque approach allows them to grow, or at least maintain, their oil and gas output to 2030 by simply adding more green stuff on top (as this means the overall carbon intensity of their energy will be reduced).[36] Similarly, if I drink a couple of extra low-alcohol beers a week, but still keep drinking the same three wee cans of full strength IPA on offer from Tesco on a Friday night, then technically the alcohol intensity of my beer consumption is lower, despite the fact I still fall asleep on the couch in front of the TV every Friday.

How quickly these companies start cutting emissions makes a big difference.[37] They could quite literally do nothing until 2049 and then cut their emissions to zero. The pathway matters, because cumulative emissions in the atmosphere matter. An end goal of zero is necessary but not sufficient.

The 'net' part of 'net zero' also gives them an escape route for cutting their emissions. Many companies appear to be relying on planting trees and suchlike to offset emissions while they continue to extract fossil fuels. In 2021, Shell released its first ever modelling scenario of how the world could reach 1.5°C. Unfortunately, its scenario for meeting the target relies on a new forest the size of Brazil to capture emissions. In 2021, a landmark court ruling in the Netherlands ordered Shell to cut its emissions by 45 per cent compared to 2019 levels by 2030 to be in line with the Paris Agreement.[38] Shell plans to appeal.

In my view, the fossil fuel industry should be massively ramping up its annual spending on carbon capture and/or preferably renewables, so that the majority of its investments are focused on these options immediately. And it has to set interim carbon reduction targets based on actual absolute emissions reductions, not intensities that are easy to hide behind.

It is entirely possible for oil and gas companies to change. Even State-owned ones. For instance, Dong Energy (the Dong stood for Danish Oil and Natural Gas) recently transitioned into one of the world's largest green energy companies, now called Orsted.[39] This transition has been a massive success: as of 2020, Orsted was worth £51 billon, double that of two years before.[40] The future needs more visionary Dongs growing across the world. I want all oil and gas companies to become Dongs.

27

What can I do about it?

This is the question I am asked about climate change most often. It's a question people would like a simple answer to. You'd like me to say: 'Do this one thing – eat more legumes – and everything will be fine.'

In reality, though, answering this question is much more complex. Let's start by dissecting the question.

Are you even responsible?

Firstly, is it even appropriate to focus on the 'I' in 'What can I do about it?'? Is it your responsibility? The correct question may well be 'What can *we* do about it?', or even 'What should *they* do about it?' In today's highly individualistic society, where we are constantly told we are important consumers with choice and agency, always positioned at the centre of the universe, our initial response is often that the answer must come from us. Yet, more often than not, I hear people say they feel helpless about effecting change on global warming: 'Does it even matter what I do?' We have all felt that way from time to time. The best place to start is in understanding whether changing our own habits is important by asking: 'Does my own carbon footprint matter?' The answer to that question is . . . no, but yes, but no.

As always, it's complicated.

Would it even be possible to halt the climate crisis if we all, together, took all the low-carbon actions we could? A campaign called Count Us In estimated in 2020 that if a billion middle-class people made various lifestyle changes to cut their carbon footprints, then global emissions could be reduced by a fifth.[1] That seems like a lot of people for not much result. I couldn't even get five people to show up to my last birthday party. (I mean, sure, I forgot to text anyone, and it was during lockdown. But still – Zoom me, fuckers.)

Often, we don't have any good climate choices to make. We're stuck. Sure, we can point to the odd better choice: eat less meat, switch your energy provider, close down the illegal coal-powered fire station you've been operating in your back garden, etc. But more often than not, certainly in the present day with currently available technological options, people make bad climate decisions because they have limited choices. It's like going to a restaurant that serves only quiche. Those bad choices are the only options available. The good ones are too expensive, inconvenient, or simply don't exist. As I already mentioned, I couldn't buy an electric car because I don't have the money and I'd have nowhere to charge it. Others have family members halfway across the world that they'd need to stop seeing if they didn't fly.

Many don't have the cash reserves to install an expensive

heat pump or solar panels. And even if they did, there are much more fun things to spend it on, like paying TV personality Ben Fogle to read you a bedtime story on Cameo for 142 nights.* Even if people want to do the right thing, reality gets in the way. We need to stop making people feel bad for simply wanting to live in today's society. The way to solve these problems is for individuals to demand solutions, which must be provided by governments and by new business models.

For example, let's think about my mate Ian. He's my most average friend, so we can assume he emits roughly the UK average emissions of 7 tonnes of CO_2e a year. He might be able to switch and give things up here and there fairly easily and inexpensively by moving back in with his sixty-five-year-old mother, having her cook for him and going on holiday with her to a caravan in St Andrews. Yet he'll probably still be left with 3 or 4 tonnes that are part of the system, or that are too expensive to deal with himself. At the moment, we need to use electricity and gas from the grid and petrol from pumps to be able to function from day to day. If we want to stop contributing to climate change entirely as individuals, then we'd essentially have to remove ourselves from society and become hermits in the woods, foraging for berries and nuts like squirrels. Getting rid of that remaining 3 or 4 tonnes requires the systems we exist in to change, and the only way you can influence that is by broadening out beyond your own individual actions and pushing for change at a societal scale. We'll come to these options later in this chapter. Plus, it is not really 'average' people that are the problem, it's wealthy high-emitters.

The very idea of a carbon footprint – whereby individuals

---

* I did the calculation.

are encouraged to take ownership of their own emissions – was famously pushed by BP as a PR technique in the 2000s to shift the onus on to consumers rather than the producers.[2] So, certainly, be sceptical about fossil fuel companies trying to push this narrative. They want you to focus on yourself, not them. Why do I feel like the Death Star would still be intact if Darth Vader had decided to use BP's PR team instead of the dark side of the force?

However, we do all need to play a role, and some more than others. Your role is a lot less than, say, Paris Hilton's, because not all consumers are equal. It is especially important that inequality is part of the climate solution debate. The UN *Emissions Gap Report 2020* found that the wealthiest 1 per cent of people emit more than double the total emissions of the poorest 50 per cent worldwide.[3] Yes, you read that right. The percentage of people in the world who earn more than $109,000 a year contribute twice as much to climate change as around 3.5 billion people. Climate change is not an excuse to get on at anyone for simply surviving by keeping themselves warm, driving to their low-paying job, or feeding their family meat if it happens to be cheaper. But that high-consuming 1 per cent are responsible for a whole lot more than what could be deemed 'essential'. Also, I'd wager a guess that those who own and run the companies that need to change their business practice in order to be more sustainable are part of that top 1 per cent. If you're a high emitter, you can and should make big cuts to your carbon footprint overnight; if everyone leads by example, and works to spread their impact to others, your actions are essentially limitless.

So those causing climate change most of all – the fossil fuel industry and the wealthy – need to take responsibility.

Unfortunately, they're the ones with more power and money to fight change and cushion themselves from the impacts.* Therefore, it is important the rest of us join together, put our differences aside, and simply keep pushing governments to change the rules and businesses to change themselves. Author and academic Dr Genevieve Guenther makes a good point when she asks who the 'we' is when we talk about how *we* are causing climate change, saying: 'To think of climate change as something that we are doing, instead of something we are being prevented from undoing, perpetuates the very ideology of the fossil fuel economy we're trying to transform.'[4]

And what does 'do' mean?

On top of all this about the 'I', there's something else to unpack in the question 'What can I do about it?' We need to expand our understanding of what constitutes the 'do'. Not knowing what we should do can lead to a lack of agency, which risks us losing momentum and falling into traps of defeatism, disengagement and doomism. We need to expand what we consider to be meaningful climate action. There are many things that you can *do* in your private, professional and political life: a list that Professor Chris Rapley likes to call 'the three Ps'.†

---

* Burn the rich! But also don't, because they've caused enough emissions already.

† On a Zoom call with him, I suggested 'the three Ps' may just be the number of times he goes to the bathroom during the night at his age. I'm not sure if the Zoom call froze at that exact moment, or if that was simply his reaction.

It is easy to define climate action simply as individual actions, like flying less or wasting less food. If we each focus on these actions and our own footprints, then, like my mate Ian, the absolute biggest effect that you can have is saving 7 tonnes of carbon. Instead, if you focus on the impact you can have on others, and therefore society, your actions are essentially limitless. It is far more effective to do what you can to cut your own footprint, but also remain part of society, changing it and helping others.

However, we also need to understand that there can be wider reasons for why these actions can be important. If you make small changes, you also send a market signal. Buying a meat-free burger increases the revenue of the meat-free burger company, enabling them to undertake research into making their products better, which, in turn, makes more people try them, which reduces the price, and so on. You're driving a new clean industry. You are also sending a signal to the old dirty producers that you no longer want their product.

When people see their neighbours have installed solar panels, they become more likely to do this themselves.[5] This so-called 'neighbourhood effect' can help spread new technologies much faster in concentrated areas than would otherwise occur. The more electric cars you see in your neighbourhood, the more it normalises having one. This is why, in the early nineties, everyone on my street had the phrase 'Suck My Balls' spray-painted on their driveway. At least that's what I thought. Turns out that it was not a sociological behavioural concept in action, but simply an unruly local teenager.* Visible actions work best,

* IT WASN'T ME, OK?!

like solar panels and electric cars. Other actions, like installing a heat pump, reducing flying, or recycling are less effective in this instance. (Apparently, it is an 'invasion of privacy' to go through your neighbours' bins to check how much they recycle.)* But do talk about your actions. Where it arises naturally, it is worth mentioning.

A less obvious thing that will help shift the dial on what is acceptable in society while simultaneously giving the finger to fossil companies, is our money. Changing your bank account and pension to one that invests in green energy can be an important step. A 2021 report from BankTrack.org and others found the world's sixty biggest commercial and investment banks have helped finance a total of $3.8 trillion in fossil fuel investments since the Paris Agreement in 2015. The largest amount came from US bank JPMorgan Chase, but Europe's worst was Barclays, which poured $145 billion into fossil fuels.[6] Websites such as Bank.Green and Switchit.Money can show whether your bank has a good or bad record on climate.[7] Triodos Bank and the Ecology Building Society both put sustainability front and centre, and provide detailed breakdowns of what they invest in. Others, like the Co-Operative Bank and Nationwide, at least explicitly state that they do not invest in fossil fuels. It also helps if you write to your old bank or pension fund to tell them why you are switching (even if all you're moving is an overdraft). Here is an example text you can use:

* Obviously I don't actually do this, as, as I said earlier, I DO NOT CARE IF YOU RECYCLE.

Dear [insert bank here],

Just because I am 'saving for a rainy day', it does not mean you should go ahead and make my days even rainier. Stop funding fossil fuels.

Yours sincerely,
[Insert your name here]

Then there are our professional lives: the institutions we work for, where we spend most of our time. What are they doing about the climate? Is my job harming the future? The good news is that it doesn't matter what you do for a living, it is still 100 per cent relevant to solving climate change. I mean, sure, you can retrain as a climate scientist, or secretly infiltrate an oil company as an employee for decades until you finally become the CEO, and then change its entire business overnight to something low-carbon but high-profit like cinema popcorn. However, these are probably more time-consuming endeavours than simply making sure that what you already do is better aligned with the world we need. We need literally all businesses to drive towards zero emissions as fast as possible.

One example is the treasurer of a local council, who, after taking a climate-training day course realised he could help the planet with his role. He was in charge of replacing twenty-two buses by the end of the year, and decided to make sure they were electric. Except he found it was too expensive to buy a fleet of electric buses. No matter. He didn't give up and went back to the drawing board. After having a good think, he set about altering the council's procurement procedure so that they didn't need to replace all the buses at once. They could be staggered, allowing most of them to be bought once they became cheaper and so they didn't end up getting stuck with lots of internal combustion buses. He is a real-life climate-action superhero. We need more of those. If the rules don't let you do the right thing, it's the rules that need changing.

Most workplaces will have at least some kind of sustainability initiative. Join it. If your workplace doesn't have this, start one. If you don't want to start one, guilt trip a colleague into doing it. This can relate to day-to-day operations, i.e. the food you eat or the travel policy. One day-to-day change forced by COVID-19 has been video conferencing, which has lowered business travel. It proved that many businesses can continue as normal while operating remotely (if, by 'normal', you mean having to tell Brenda she's muted herself for the seven thousandth time that day). Sustainability initiatives can also relate to work the business undertakes. For instance, architects can help reduce the amount of cement and steel required for building designs.

Now, I don't mean to get political but, well, get political. That is to say, vote for strong climate action. And if the party you normally vote for doesn't include strong action in their manifesto, tell them to get with the programme. Climate science is apolitical. As Professor Katharine Hayhoe often says, 'A thermometer isn't liberal or conservative; it doesn't give you a different number depending on whether you vote left or right'.[8] Vote like your future and your children's futures depend on it. Because they do. Email your Member of Parliament again and again about climate change and tell them that you care about it. MPs often say that they don't feel they can take action because none of their constituents contact them about climate change. This is a remarkable position to hold – 'I would have helped avert an environmental apocalypse, but nobody emailed me about it' – but shows how difficult it can be to solve systemic problems like this in a democracy.

Despite this apparent lack of emailing, people actually do clearly support action. Recent surveys have consistently shown increased concern about climate over the last years. One UK poll showed that 39 per cent of people were very or extremely worried about climate change in 2020, up from 25 per cent in 2016.[9] And a UK government survey in March 2019 found that 80 per cent of people were either fairly (45 per cent) or very (35 per cent) concerned about climate change.[10] This growing concern has been seen across towns and villages, too, not just big cities.[11] So if you care, make sure that those in charge hear you.

This is action

That brings us to the important role of protest. There have been two notable examples of this in the last few years.

In late 2018, a new protest group popped up I had never heard of before. They're called Extinction Rebellion (sometimes shortened to 'XR'). You're probably aware of them, but if not, I like to describe them as the BrewDog of climate activism. Good branding. Appeal to a certain type. They rocked up out of nowhere and started taking 'direct action'. This meant they glued themselves to buildings and blocked bridges; they even held a 'die-in' at the Natural History Museum, which meant that for thirty minutes, a T-Rex was forced to become a climate protester.

At first, I personally found their approach tough to get on board with. I had worked on climate change for eleven years, well before it was cool. Hot. Whatever! And I'll be honest, no one cared. But I kept going. Then, all of a sudden, all these people were getting into climate change. It felt like when others gets into a band you like. As a teenager, I was a huge fan of the band Biffy Clyro; I had their first four albums, I went to their gigs. And then my mum bought their fifth album and I was like, 'That's me done!'

But I decided not to just go on my first judgemental instincts, and instead try to understand Extinction Rebellion's approach.

Their main goal has been getting themselves arrested. And I can understand why. With climate change, the solutions are never commensurate with the scale of the problem. We are being told the world is burning and so we'd better change our light bulbs. Which is like saying that you're entering a war zone, so you'd better put on some knee pads. Getting arrested signals that you're willing to do something big (although, as others rightly pointed out, getting arrested in this way is a white privilege). So I began to understand their motivation, even if I sometimes disagreed with their methods. And people complain they are disruptive, which is true. But once, during the summer of 2019, I couldn't get out of London because the railway tracks were melting, and that was quite disruptive too. So maybe disruption is coming whether you like it or not. Extinction Rebellion's work has led to many declarations of climate emergencies and a rise in media coverage and awareness of the topic. People are discussing climate change. Now, obviously, you don't have to join XR. I haven't. My advice is simply to find your own group, as by working with others, you can achieve so much more.

The other major climate protest movement of the past few years is the school strikes. In August 2018, one solitary Swedish teenager sat outside her country's parliament asking for change.

Eight months later, in March 2019, an estimated 1.6 million children in over 125 countries walked out of school.[12] My only issue is that they called it Youth Strike for Climate and not the Minors' Strike.

At the time, UK Prime Minister Theresa May* said that the school time they were missing: 'is crucial for young people precisely so that they can develop into the top scientists, engineers and advocates we need to help tackle this problem.' I went along to one of the strikes in London. You do not need to worry about the education of the kids on those marches: they were all absolute nerds. I mean, seriously, Greta skips school to study and talk about science. The only children affected were the school bullies turning up on a Friday with nobody to wedgie. That so many of the leaders of this movement are young girls is wonderful, and interesting given a recent study that showed that daughters are particularly influential in raising concern about climate change in their parents, especially fathers.[13] That the movement has also spread to many places in the Global South is a testament to the clarity of the message and the inclusive appeal of a powerful act. It feels like an entire global generation of all types speaking together with one voice.

Ms Thunberg responded to May in her classic understated manner, saying that politicians don't listen to the scientists, engineers and advocates anyway, so what's the point. As one of those scientists, I wholeheartedly agree. I find it necessary to spend my evenings performing comedy about climate change just to feel as if I am doing all I can to help people understand what is happening. If I thought that governments would simply

* Remember her? Long time ago, eh?

read my scientific papers and that would be that, this book would not exist.

I believe the success of the school strikes has been profound for two reasons. Firstly, like the risk of getting arrested for XR protesters, the act of striking is a big deal because for kids, their education is all the power they have. And for young people, the focus on system change is a natural one. They have little agency, they cannot vote, they are not the wealthy high emitters who are in control. They are dependent upon adults and upon the system adults have constructed. Yet it is them who will deal with the consequences of this system throughout their lives.

Many people have told me that seeing those school strikes has helped give them hope for the future. I'd say, if you are waiting for others, especially children, to give you hope about the future, then you are in the wrong game. Sure, you may get some nice feelings by watching a twelve-year-old holding a placard, but the point is that action breeds hope. The very reason they are on the street is because you are doing nothing. They should not be giving you hope. You should be giving them hope. Do something about it yourself.

For me, a terrific thing to come out of all the protests was the formation of a citizens' Climate Assembly. The government accepted a proposal to bring ordinary people centre stage in the debate about what to do about the climate. One hundred and eight people from all over Britain were randomly chosen to spend four weekends learning about climate from experts and discussing it among themselves. Many are on record saying it changed their outlook on the subject and inspired them to make life changes.[14] One is on record saying the quality of the biscuits supplied at the seminars was poor.

Having ordinary people going on a journey to learn for

themselves what needs to happen and why is inspiring and necessary. It cannot be overemphasised how important it is to feel part of the decision-making process. Telling someone they need to pay more to take a flight is a hard sell. But if they go on a journey to understand why flying is so polluting and are given options to solve it and reach the conclusion themselves that we should be paying more for air travel, it gets easier. Sure, it's a pretty labour-intensive method. But it works.

Many ordinary people are already getting involved in changes at the community level. Community is perhaps the missing link in the discussion about individual lifestyle choices vs massive system change and big government. At local levels, it is easier to work together to build the sort of future that you actually want in your neighbourhood. In the UK, over 200 local authorities have declared climate emergencies.[15] That implies 86 per cent of UK citizens live within local authorities that have declared a climate emergency.[16]

There are countless inspiring examples of small communities taking action. The Goodwin Development Trust in Hull have renovated abandoned homes to *Passivhaus* standard to provide high-quality low-carbon social housing.[17] Repair cafés are popping up around the country, with about 150 across the UK.[18] The whole village of Swaffham Prior is moving from oil heating to district renewable heating.[19] A benefit fund from the Isle of Mull's community-owned hydro energy has helped finance new equipment for a local playpark, shelves for the library and a summer camp for primary pupils.[20] The Plymouth Energy Community tackles both climate change and local fuel poverty. It built a community-owned solar farm that provides 1,000 homes with clean energy and has supported over 2,800 households.[21]

Perhaps the most important piece of advice is that climate action that works for you will depend upon your unique situation and traits. It has to be a journey of your own to feel meaningful. Dr Kris De Meyer says to ask yourself: 'What can I uniquely do that other people may not be able to do in the same way?' This can be a much more rewarding way for people to start thinking, he adds. For me, it was combining my career with my semi-professional hobby to bring a unique perspective to communicating the problem. That's my climate action. I get an inordinate amount of satisfaction from communicating the subject in a fun way, and it is also helping people take more action than I could alone.*

The morning after my first performance at the Edinburgh Fringe in 2017, I got an email from a local chap called Stuart telling me he had heard me on sports radio promoting my show. He'd come along to see it, and found what I had to say thought-provoking. Then he said he'd just switched electricity provider – something I mentioned doing in the show – and was really happy, as he'd saved money and gone renewable. This sparked an idea. I decided to include my energy referral in

---

* I introduced Ian to a friend of mine, Lauren, who also works on climate change, and they have gone on a few dates. He wouldn't listen to me about the subject but now that it's coming from someone he fancies, he won't bloody shut up about it. He's planning on buying an EV next year he says. I count this matchmaking as even more climate action. However, it is quite annoying and does mean I now hear facts I have told him repeated back to me – like how to properly load the dishwasher. Worth it though. Last weekend, they went for a tandem ride together. Lauren has even taught him to mellow out. He's now the best man at his ex-wife and Vincent the CrossFit instructor's wedding.

future performances in order to track my influence.[*] Last year, according to our energy provider this helped save 141 tonnes of CO_2e.[†] Which, even if the 58 tonnes per child was correct (it's not), is enough to offset almost three Oscars – nearly a Daniel-Day Lewis.[‡] Although I must be clear, this is a joke. Offsetting is a nonsense and the idea of offsetting a baby is obviously stupid.

In 2018, a young couple who'd seen my show tweeted to say it was the most expensive comedy show they'd ever attended, because they sold their car afterwards and bought an electric one. In 2019, a family who saw me perform mentioned they'd skipped their annual family holiday abroad and stayed in Scotland. (They also sent me some socks in the post after noticing I had holes in mine while on stage.) Point is, we can have a much larger influence when we think beyond our own personal emissions.

Recently, I've had teenagers come up to me after shows to ask me what they should do with their lives to help stop climate change. I never for a moment thought when writing comedy that I'd also become a careers advisor. The first few times, I didn't have a prepared answer, but would mumble something about finding out what it is you are good at, what you enjoy, and combining that with working somewhere you can make the biggest difference. Turns out my instincts were right. Kris De Meyer adds that discovering how each of us are uniquely placed to act on climate change is 'the psychological problem of the

---

* And to rake in that sweet, sweet discount on energy bills for a while.

† According to energy provider Bulb, the average member saved 1.31 tonnes of CO_2 last year. Forty-nine referrals is 64.19 tonnes of CO_2.

‡ Lucky we didn't have triplets.

moment'. It's not about telling people what to do, he says, but helping them to discover how to do what they already want to do. Rather than preprepared menus of climate actions to order, we need recipe books to help people create their own actions.

Overall, if your emissions are small, then don't sweat it. Focus on making your voice heard and think about where you might have the biggest impact, whether that's in your career, political actions or your personal life.

If your emissions are big, then look in the mirror. Even then, working towards changing the system is often going to have more impact than stressing about your own emissions. Like, Emma Thompson flying in from LA to London for an Extinction Rebellion protest is pretty stupid, sure. But it is probably better than her not flying while never engaging in protest.

Broadly, our actions can be thought of in three parts:

1. **Our individual actions.** We need to understand there are more actions than the obvious ones (e.g. what about pensions, banking, our workplaces) and better understand which of these have the biggest impacts.
2. **Peer and community effects.** We need to be aware that our actions are part of society, our community, and can influence others.
3. **System and societal change.** We need to shift the focus from being simply consumers to being active citizens. We need to make changing the systems we live in an essential part of climate action. This may be political or local action, or finding your own climate action in another way, e.g. a global warming themed juggling act.

Considering all three of these aspects together can help you identify meaningful action towards a better future. If you cannot do much with regards to (1), then don't worry; focus on (3). It has the potential for far greater impact anyway. Regardless, building a better future requires what the writer Mary Annaïse Heglar calls not 'climate action' but 'climate commitment'.[22]

28

Nine months

I make decisions that feel right for me and my baby

Once again, I am sitting down to write while watching him. He's wriggling a lot, and the baby monitor tells me it is hot in his room. We're finding it hard to get baby sleeping bags in low enough togs because the UK has not usually been this warm in the summer, and we live in a 110-year-old house that was designed for a previous, stable climate.

If this were a novel, the story would tie up nicely right about now. This would be the end. Our global warming journey, from pregnancy to birth to baby, would reach its climatic conclusion. It's not. This is the beginning. My parenting journey has just started, and so has our collective journey to stop altering the planet.

And there are similarities with these journeys – childrearing and climate action. Both can focus too much on a fear of the future and a concern about what *might* happen. I do, on occasion, find myself playing out worst-case scenarios in my head. *What happens if he bangs his head or simply stops breathing?* Yet I think the new father part of me, Matt, can learn from the climate academic Matthew on this one. Fear can be paralysing when what we really need more than anything are acts and deeds. Worry is no substitute for action and cautious optimism. If you have an overwhelming desire to do right and be a good dad, then that is all you can do. Commit, every day. And know that it will be messy, because life is messy, and that you can't get everything right. That you will make mistakes; you're only human.

I will try to instil optimism and positivity in my son and hope he will benefit from happiness and joy. Much like the climate movement can. Yet he must be aware that nothing is handed to you on a plate; that actions are more important than words.* I hope I give him the confidence to tackle the world head-on, without the over-confidence that many of those in charge appear to have that they don't have to work for their privilege.

Also, parenting and climate are both really, really hard. Jesus, they are exhausting, chaotic and messy but worth the effort. Many times, things will happen that make you want to give up. It can be tough to keep going. I think the academic part of me can learn from the dad part here. Every single morning, I am woken by Oscar feeding in bed next to me, and then, when he has finished gulping milk, he turns triumphantly towards me to climb directly on to my face and play, smiling while pulling my hair and putting his fingers in my mouth. That's what makes it all fine. His smile. I need it so much; it keeps me going. Each and every day. I've always really struggled to wake up and get

---

* I realise that this is a book and therefore almost 100 per cent words. But, in my defence, I had to *write* the book, and that in itself is an action!

out of bed in the morning. I've felt a malaise my entire adult life; not any more. What I would say is you have to find the things that keep you going and do them often: discover joy in the small things and celebrate the good times.

And I can assure you that we still have time to avert the worst, just like we've managed to baby-proof our living room and buy a playpen so Oscar stops trying to climb up the radiator. The worst horror stories about our planet need not come to pass. And every parenting journey is different, just as what each of us can bring to the climate movement is different. No two journeys are the same. You are unique, and so is what you can do. People give you all sorts of advice. 'Never let them nap after 3.14pm', 'Hold them upside down and play them Pavarotti, it always, without fail, calms them down', 'Our baby slept through the night from the day they were born, but only because they wore a cravat', and other such unlikely guff. Like climate action, it is about doing it yourself. Finding out what works for you and your situation.

Also, you have to be in them both for the long haul, or so I'm told. These are big commitments. These two things will define the next decades of my life and those of the people I love. I mean, sure, occasionally I think I should have got a job working as a childless YouTuber who yachts about the world, but then that's not really a job, is it? Most of the time, I realise how much purpose both my son and working on climate change give to my life on a daily basis. Hashtag blessed. Yet, already, time has flown by. I'm trying to savour every precious moment with Oscar and use each one wisely. I have heard people say that with parenting, the days are long but the years pass quickly. I think it is the same with climate change. There, time may be our biggest constraint; it is passing quickly too. 2050 may be the year after the *Blade Runner* sequel is set, but there are many high-carbon decisions being made today that will still be highly relevant then, like investments in new power plants and steel mills that will still be operating in 2050. We can't start in 2049. The story needs to start today, and get better before then.

Oscar is nine months old now. He's started ticking off milestones left, right and centre. He has seven teeth. A couple of weeks ago, he only had three – it's weird, as they keep coming in in odd numbers. He started eating solid foods a few months ago. I know now where the phrase 'egg on your face' comes from. And, as most of his food ends up on the floor, I can't help thinking he might be responsible for more food-waste-related climate change than I expected. At least the portions are small. At seven months, he started standing up in his cot. Then there were a few hellish days where he couldn't get back down, so he'd wake up crying, standing up and staying up until one of us talked him back down . . . Well. Picked him up and put him back down. Finally, after a couple of sleep-deprived nights, he worked out how to sit back down as if he'd been doing it his whole life. That same week, the earth hit a milestone of its own. Carbon dioxide concentrations in the atmosphere passed a threshold whereby they have now increased by 50 per cent since the Industrial Revolution: a rise as unprecedented and dangerous to humanity as Ed Sheeran's popularity. To put that in context, the last time the planet had concentrations this high was probably about three million years ago. Back then, the Arctic would have had forests instead of ice, with crocodiles swimming nearby, and the Sahara would have had vegetation. Sea levels would have been 30 to 50 feet higher, meaning New York and Miami would be sleeping with the fishes/Kevin Costners. It may take a while, but this is where we are eventually heading unless we bring emissions down to zero ASAP.

How do I sum up what I've learned? If I was to give you one opinion, it would be this.

If you want to have children, then have children. Climate change should not factor into this. If you don't want to have children because you're worried about climate change, then of course that is completely your right to decide. But you don't *have to* have no children. But it's not really about having kids or not. It is about bringing crucial tools of learning, compassion and understanding into whatever are the right climate choices that apply to you. For me it is my son, for you it could be what you buy, studying, your job, protesting or (heaven forbid) becoming a politician.

I wrote this book while still doing my job, moving to a new house and having a child, all during a pandemic. Would I advise it? Well, no, probably not. If this book makes any sort of sense whatsoever, it will be a miracle. But life still goes on. During a pandemic. During a climate crisis. People get married, die and are born. People have always continued having kids during difficult times. People continued having kids during World War Two, during the Plague, and during season eight of *Game of Thrones*. As I watched the early weeks of the pandemic, it felt so similar to a sped-up version of the climate crisis. Like COVID, the longer we hold off taking real action, the more painful, deep and drawn-out the effects are on our health, on the economy, and on leading normal lives.

I hope this book has been helpful in some way. It has been my pep talk about building a better world with zero emissions, clean air and a stable climate. A safe world for our children. One built on fairness for all. But this is not a movie. I mean, of course not – it's a book. Although I am very happy to sell the rights to Disney, and maybe Baby Yoda can play my son. But what I mean is, it's not a fiction book. As I said, the story has only just begun. There are many climate futures the world can still take. Dr Lydia Messling, a climate justice and adaptation expert, says: 'I describe climate projections as being like different storylines, different plot lines, and you're going to be an actor in them. They're not predictions, they are projections [. . .] and depending on which storyline we go for, you'll get a different part. So choose which part you want.'

My point is that you'd better go do some sort of climate action after reading this, otherwise I have missed many days of the first year of my son's life writing this book for no reason whatsoever. So, do it for me. And if not for me, then do it for him. And if not for him, do it for people that need help more than him and me, and your community. And if not for them, just do it for your loved ones. And if you don't have loved ones, then do it for yourself. And if you hate yourself, then do it for David Attenborough. We all love him. I don't really care why you do it. Find your own action. Find your own reason.

At the start of the book, I jokingly described climate change as the world's shittiest *Choose Your Own Adventure* game. And it's true. Which climate future the world ends up with is dependent upon all of us. You aren't simply reading the story, you are a part of it. So the outcome doesn't need to be shit at all, you can decide what you do. Sure, the stakes are slightly higher than Spot the Dog losing his teddy bear, they are higher than really anyone can cope with alone, but you are not alone, none of us are. That's why it's so important that you take part, because we have to work together. You cannot put this book down and simply get back on with your life as if nothing has happened, satisfied with a nourishing narrative. The next action you take after you finish reading the last words on this page will be the next part of this story. There is no ending. Not now. Your move.

Acknowledgements

That was my book. The first person I'd like to thank is you for taking the time to read it. I did think about doing a 'no thank you section' where I just listed deniers and others that hinder climate action but best not to give them the limelight. Let's keep it positive and actually thank all the people that helped make this book a reality.

I would like to thank Iona and Rob at Avalon for all their hard work behind the scenes. And to Richard Roper at Headline Publishing for both making this happen in the first place and keeping me going during a pandemic and a birth. Your positivity helped motivate me when I thought I could not go on.

I really need to thank a bunch of lovely people who gave notes, jokes, and fact-checking to make sure the book was semi-coherent, funny and accurate: James Kirk, Jocelyn Timperley, Tom Baker, James Rowland, Eleanor Morton, Charlie Dinkin, Tom Parry, Ben Clark, Freya Garry, Kirsten Lees, Ken Rice, Mark Maslin, Will McDowall, Christophe

McGlade, Martin Croser, Steve Dunne, Stuart Laws, Keith Alexander, Sam Hill, Leo Murray.

A number of experts who gave their time and input, many of whom I knew and am grateful to, and others who I contacted out of the blue about a comedy book about climate change: Kris De Meyer, Lydia Messling, Chris Rapley, Leo Hickman, Kimberley Nicholas, Jillian Anable, James Beard, Brianna Craft, Richard Lowes, Faye Wade, Tadj Oreszczyn, Mat Hope, Richard Black, Stuart Capstick, David Harkin, Giulio Mattioli, James Price, Steve Pye, Isabella Kaminski, Dominic Roser, Steven Kirk, Andy Carr. So sorry if I have forgotten anyone. I'm writing this at half past midnight and need to get to bed as I will likely be woken up at 6am.

And to all the supportive people I work with at University College London.

A big appreciation to our new neighbours who have been incredibly welcoming to us during a time of much change. I will hopefully no longer be sat at the window at midnight any more.

Thank you to all my family and close friends for your love and support over the years. There were some who passed away during the last year and a half and, because of COVID, I grieved for you while sat in front of a computer screen. I miss you terribly – JB and SW.

And the largest thanks of all goes to my wife. To say this would not be possible without her is an understatement of monumental proportions. She sacrificed so much during a time when sacrifices were hard to make. I cannot begin to tell you how much your support, input and patience has meant, and your kindness will stay with me forever.

References

Chapter 1

1 Duggal, D. 'Oxford reveals Word of the Year 2019: Here's why we should be very, very concerned' *Economic Times*, 5 December 2019.

2 Clayton, S., Manning, C. M., Krygsman, K., and Speiser, M. *Mental Health and Our Changing Climate: Impacts, Implications, and Guidance.* American Psychological Association and ecoAmerica, March 2017.

3 Doherty, S. 'The activists going on "birth strike" to protest climate change'. *Vice*, 13 March 2019.

Chapter 3

1 Hoesung, L. 'High-level segment of COP24, Tuesday 11 December 2018, Statement by IPCC Chair Hoesung Lee'. Intergovernmental Panel on Climate Change, 11 December 2018.

Chapter 4

1 Rapley, C. and MacMillan, D., *2071: The World We'll Leave Our Grandchildren*, John Murray, 2015

2 Brazil, R. 'Eunice Foote: The mother of climate change'. www.chemistryworld.com, 20 April 2020.

3 Kamiya, G. 'Factcheck: What is the carbon footprint on streaming video on Netflix?' www.carbonbrief.orfg, 24 February 2020.

4 Jones, N. 'How the world passed a carbon threshold and why it matters'. Yale School of the Environment, 26 January 2017.

5 Lindsey, R. 'Climate change: atmospheric carbon dioxide'. www.climate.gov, 14 August 2020.

6 Rohde, R. *Global Temperature Report for 2020*. Berkeley Earth, 13 January 2021.

7 Met Office. 'Top ten UK's hottest years all since 2002'. www.metoffice.gov.uk, 31 July 2019.

8 Ramsayer, K. '2020 Arctic Sea Ice Minimum at Second Lowest on Record'. NASA's Goddard Space Flight Center. www.climate.nasa.gov, 21 September 2020.

9 Maslin, M. *How To Save Our Planet: The Facts*. Penguin, 2021.

10 Maslin, M. *Climate Change: A Very Short Introduction*. Oxford University Press, 2014. p.18

11 Myhre, G., D. Shindell, F.-M. Bréon, W. Collins, J. Fuglestvedt, J. Huang, D. Koch, J. F. Lamarque, D. Lee, B. Mendoza, T. Nakajima, A. Robock, G. Stephens, T. Takemura and H. Zhan. 'Anthropogenic and Natural Radiative Forcing.' Climate Change 2013: *The Physical Science Basis. Contribution of Working Group I to the Fifth Assessment Report of the Intergovernmental Panel*

on *Climate Change*. Edited by Stocker, T.F., D. Qin, G.K. Plattner, M. Tignor, S. K. Allen, J. Boschung, A. Nauels, Y. Xia, V. Bex and P. M. Midgley. Cambridge University Press, 2013.

12 Hausfather, Z. 'Analysis: Why scientists think 100% of global warming is due to humans'. "http://www.carbonbrief.org" www.carbonbrief.org, 13 December 2017.

Chapter 5

1 Wynes, S. and Nicholas, K., 'The climate mitigation gap: education and government recommendations miss the most effective individual actions', *Environmental Research Letters,* Vol. 12, No. 7, 12 July 2017.

2 OECD, 'Greenhouse gas emissions', data extracted on 21 Aug 2021, 13:09 UTC (GMT) from OECD.Stat.

3 Wynes, S. and Nicholas, K., 2017.

4 Ivanova, D., and Wood, R., 'The unequal distribution of household carbon footprints in Europe and its link to sustainability', *Global Sustainability*, 3, E18, doi:10.1017/sus.2020.12, 6 July 2020.

Chapter 6

1 Rustemeyer, N. and Howells, M. 'Excess Mortality in England during the 2019 Summer Heatwaves'. *Climate* 2021, 9, 14. https://doi.org/10.3390/cli9010014.

2 Carrington, D. 'Heatwaves in 2019 led to almost 900 extra deaths in England'. *Guardian*, 7 January 2020.

3 CCC. 'Independent Assessment of UK Climate Risk (CCRA3)'. www.ukclimaterisk.org/. 2021

4 Timperley, J. 'In-depth: How the UK plans to adapt to climate change'. www.carbonbrief.org, 26 July 2018.

5 Watts, N., Amaan, M., Arnell, N., Ayeb-Karlsson, S., Beagley, J., Belesova, K., *et al*. 'The 2020 Report of The Lancet Countdown on health and climate change: responding to converging crises'. *The Lancet*, Vol. 397, Issue 10269, 9 January 2021.

6 Watts, J. 'Canadian inferno: northern heat exceeds worst-case climate models'. *Guardian*, 2 July 2021.

7 Polderman, J. 'Lytton's mayor: "Where many buildings stood is now simply charred earth"'. *Guardian*, 8 July 2021.

8 Vaughan, A. 'Climate change made North American heatwave 150 times more likely'. *New Scientist*, 7 July 2021.

9 Brown, S. 'Future changes in heatwave severity, duration and frequency due to climate change for the most populous cities'. *Weather and Climate Extremes*, Vol. 30, 100278, December 2020.

10 NASA Climate Kids. 'What is a heat island?' www.climatekids.nasa.gov.

11 Askew, A. and Tandon, A. 'Met Office: The UK's record-breaking August 2020 heatwave'. www.carbonbrief.org, 10 September 2020.

12 BBC News. '"Highest temperature on Earth" as Death Valley, US, hits 54.4°C'. www.bbc.co.uk/news, 17 August 2020.

13 Henson, R. *The Thinking Person's Guide to Climate Change*. American Meteorological Society, 2019. p. 65.

14 BBC News. '"Highest temperature on Earth"', 17 August 2020.

15 Masters, J. 'Death Valley, California, breaks the all-time world heat record for the second year in a row'. www.yaleclimateconnections.org, 12 July 2021.

16 World Meteorological Organization. '2020 on track to be

one of three warmest years on record'. www.public.wmo.int, 2 December 2020.

17 Askew, A. and Tandon, A. 'The UK's record-breaking August 2020 heatwave'. Carbon Brief, 10 September 2020.

18 World Meteorological Organization. 'Reported new record temperature of 38°C north of Arctic Circle'. www.public.wmo.int, 23 June 2020.

19 Dunne, D. 'Siberia's 2020 heatwave made '600 times more likely' by climate change'. Carbon Brief, 15 July 2020.

20 NOAA. 'It's official: July was Earth's hottest month on record'. www.noaa.gov. 13 August 2021.

21 Dunne, D. 'Climate change made Europe's 2019 record heatwave up to "100 times more likely"'. www.carbonbrief.org, 2 August 2019.

22 *New Scientist* and Press Association. 'UK temperatures broke records in 2019 as climate change took hold'. *New Scientist*, 30 July 2020.

23 Roads: Held, A. 'Melting roads and runny roofs: Heat scorches the Northern hemisphere'. www.npr.otg, 5 July 2018.

Railways: Spector, D. 'Why extreme heat turns train tracks into spaghetti'. www.businessinsider.com, 25 July 2019.

Power: Prociv, K. and Lozano, A. V. 'Sweltering heat is shattering records, triggering power outages across California'. www.nbcnews.com, 14 August 2020.

Planes: Hope, A. 'It's too hot in the southwest for planes to fly – here's why'. *Condé Nast Traveler*, 20 June 2017.

24 Matthews, T. and Raymond, C. 'Global warming now pushing heat into territory humans cannot tolerate'. The Conversation, 20 May 2020.

25 Maslin, M. 'Will three billion people really live in temperatures as hot as the Sahara by 2070?'. The Conversation, 6 May 2020.

26 Watts, N., *et al*. 'The 2020 Report of The Lancet Countdown on health and climate change'. 2021.

27 Mitchell, E. 'Pentagon declares climate change a "national security issue"'. *The Hill*. 27 January 2021.

28 Trenberth, K., Dai, A., van der Schrier, G., Jones, P., Barichivich, J., Briffa, K. and Sheffield, J. 'Global warming and changes in drought'. *Nature Climate Change*, Vol. 4, 17–22, 20 December 2013.

29 Chen, J. and Mueller, V. 'Climate change is making soils saltier, forcing many farmers to find new livelihoods'. www.theconversation.com, 29 November 2018.

30 Chen. J. and Mueller, V. 'Coastal climate change, soil salinity and human migration in Bangladesh'. *Nature Climate Change,* Vol. 8, 981–985, 22 October 2018.

31 Deamer, K. 'California's long drought has killed 100 million trees'. www.livescience.com, 7 December 2016.

32 Gudmundsson, L. and Seneviratne, S. I. 'Anthropogenic climate change affects meteorological drought risk in Europe'. *Environmental Research Letters*, Vol. 11, no. 4, 7 April 2016.

33 Carrington, D. 'England could run short of water within 25 years. *Guardian*, 18 March 2019.

34 Readfern, G. '2019 was Australia's hottest year on record – 1.5°C above average temperature'. *Guardian*, 1 January 2020; Readfern, G. 'Australia records its hottest day ever – one day after previous record'. *Guardian*, 19 December 2019.

35 Carlsberg Group. 'Carlsberg Group announces innovative

partnership to protect shared water resources in India'. www.carlsberggroup.com, 24 November 2020.

36 Energy Saving Trust, 'At Home With Water', 2013.

37 Mahr, K. 'How Cape Town was saved from running out of water'. *Guardian*, 2 May 2018.

38 Torrent Tucker, D. 'In a warming world, Cape Town's "Day Zero" drought won't be an anomaly, Stanford researcher says'. www.news.stanford.edu, 9 November 2020.

39 World Health Organization. 'Fact sheet: Drinking-water'. www.who.int, 14 June 2019.

40 McGrath, M. 'Climate change: Warming made UK heatwave 30 times more likely'. www.bbc.co.uk/news, 6 December 2018.

41 Xie, W., Xiong, W., Pan, J., Ali, T., Cui, Q., Guan, D., Meng, J., Mueller, N., Lin, E. and Davis, S., 'Decreases in global beer supply due to extreme drought and heat'. *Nature Plants,* Vol. 4, 964–73, 15 October 2018.

42 Maslin, M. 2014. p. 93.

43 Lenoir, J., Bertrand, R., Comte, L., *et al*. 'Species better track climate warming in the oceans than on land'. *Nature Ecololgy and Evolution* 4, 1044–1059 (2020), https://doi.org/10.1038/s41559-020-1198-2.

44 habel, J. C., Rödder, D., Schmitt, T., Gros, P., *et al*. 'Climate change drives mountain butterflies towards the summits'. *Scientific Reports* 11, 14382 (2021). https://doi.org/10.1038/s41598-021-93826-0.

45 BBC News. 'Bramble Cay melomys: Climate change-ravaged rodent listed as extinct'. www.bbc.co.uk/news, 20 February 2019.

46 Román-Palacios, C. Wiens, J. J. 'Recent responses to climate change reveal the drivers of species extinction and survival'. *PNAS*. 117 (8) 4211-4217, February, 2020, DOI: 10.1073/pnas.1913007117.

Chapter 7

1 Cheng, L., *et al*. 'Upper Ocean Temperatures Hit Record High in 2020'. *Advances in Atmospheric Sciences*, 38, 523–30, 2021.

2 Maslin, M. 2014.

3 Slater, T., Lawrence, I., Otosake, I., Shepherd, A., Gourmelen, N., Jakob, L., Tepes, P., Gilbert, L. and Nienow, P. 'Review article: Earth's ice imbalance'. *The Cryosphere,* Vol. 15, 233–46, 2021.

4 Henson, R. 2019.

5 IPCC. 'Special report: Global warming of 1.5°C'. www.ipcc.ch.

6 Lindsey, R. 'Climate change: Global sea level'. NOAA, www.climate.gov, 25 January 2021.

7 Henson, R. 2019. p. 113.

8 Kater, I. 'Mass starvation of reindeer linked to climate change and habitat loss'. *The Conversation*, 6 August 2019.

9 Zekollari, H., Huss, M., and Farinotti, D. 'Modelling the future evolution of glaciers in the European Alps under the EURO-CORDEX RCM ensemble', *The Cryosphere*, 13, 1125–46, https://doi.org/10.5194/tc-13-1125-2019, 2019.

10 Gobiet, A., *et al*. '21st century climate change in the European Alps – A review'. *Science of the Total Environment*, Vol. 493, pp. 1138–51, 15 September 2014

11 Lindsey, R. 2021.

12 Sample, I. 'Sea level rise doubles in 150 years'. *Guardian*, 25 November 2005.

13 Voosen, P. 'Seas are rising faster than ever'. www.sciencemag.org, 18 November 2020.

14 Lindsey, R. 2021.

15 Lindsey, R. 2021.

16 Stocker, T. F., Qin, D., Plattner, G. K., Tignor, M., Allen, S. K., Boschung, J., Nauels, A., Xia, Y., Bex, V. and Midgley, P. M. (eds.).. *Climate Change 2013: The physical science basis. Contribution of Working Group I to the Fifth Assessment Report of the Intergovernmental Panel on Climate Change*, Cambridge University Press, 2013.].

17 Watts, J. 'Sea levels could rise more than a metre by 2100, experts say', *Guardian*, 8 May 2020.

18 Oppenheimer, M., Glavovic B. C., Hinkel, J., van de Wal, R., Magnan, A. K., Abd-Elgawad, A., Cai, R., Cifuentes-Jara, M., DeConto, R. M., Ghosh, T., Hay, J., Isla, F., Marzeion, B., Meyssignac, B., and Sebesvari, Z., '2019: Sea Level Rise and Implications for Low-Lying Islands, Coasts and Communities. In: *IPCC Special Report on the Ocean and Cryosphere in a Changing Climate*. Pörtner, H.-O., Roberts, D.C., Masson-Delmotte, V., Zhai, P., Tignor, M., Poloczanska, F., Mintenbeck, K., Alegría, A., Nicolai, M., Okem, A., Petzold, J., Rama, B,., Weyer, D.C. (eds.).

19 Strauss, B. H., Kulp, S. and Levermann, A. *Mapping Choices: Carbon, Climate, and Rising Seas, Our Global Legacy*. Climate Central, November 2015. pp. 1–38.

20 Holder, J., Kommenda, N. and Watts, J. 'The three-degree world: the cities that will be drowned by global warming'. *Guardian*, 3 November 2017.

21 Letman, J. 'Rising sea levels give island nation a stark choice: relocate or elevate'. *National Geographic*, 19 November 2018.

22 Wall, T. '"This is a wake-up call": the villagers who could be Britain's first climate refugees'. *Guardian*, 18 May 2019.

23 Watts, J. 'Lincolnshire's coast and farms will sink with 3°C of warming'. *Guardian*, 3 November 2017.

24 Baker, H. 'Baby sharks are born scrawny and sick because of climate change'. www.livescience.com, 13 January 2021.

25 Carrington, D. 'Climate crisis pushing great white sharks into new waters'. *Guardian*, 9 February 2021.

26 Briggs, H. 'Squid may become favourite UK meal as seas become warmer'. *BBC News,* 12 December 2016.

27 National Oceanic and Atmospheric Administration. 'Ocean acidification', www.noaa.gov.

28 Henson, R. 2019. p. 178.

29 Carbon Brief. 'The impacts of climate change at 1.5°C, 2°C and beyond'. www.carbonbrief.org, 4 October 2018.

Chapter 8

1 Wynes and Nicholas. 2017.

Chapter 9

1 Neslen, A. 'Flood disasters more than double across Europe in 35 years'. *Guardian,* 19 January 2017.

2 Henson, R. 2019. p. 81.

3 Neslen, A. 2017.

4 Kahraman, A., Kendon, E. J., Chan, S. C., and Fowler, H. J. 'Quasi-stationary intense rainstorms spread across Europe under climate change'. Geophysical Research Letters, 48, e2020GL092361, 2021, https://doi.org/10.1029/2020GL092361.

5 McGrath, M. 'Climate change: Europe's extreme rains made more likely by humans'. *BBC News*, 23 August 2021.

6 Katzenberger, A., Schewe, J., Pongratz, J., and Levermann, A. 'Robust increase of Indian monsoon rainfall and its variability under future warming in CMIP-6 models'. *Earth System Dynamics*, 2021. DOI: 10.5194/esd-2020-80.

7 BBC News. 'Japan landslide: 20 people missing in Atami city'. www.bbc.co.uk/news, 4 July 2021.

8 PA Media. 'Met Office confirms UK had its wettest day on record after Storm Alex'. *Guardian*, 16 October 2020.

9 Otto, F., *et al*. 'Climate change increases the probability of heavy rains in Northern England/Southern Scotland like those of Storm Desmond – a real-time event attribution revisited'. *Environmental Research Letters*, Vol. 13, 29 January 2018.

10 Energy & Climate Intelligence Unit. 'Briefing: Flood risk and the UK'. www.eciu.net.

11 Climate Change Committee. 'UK Climate Change Risk Assessment 2017: Evidence Report', 2016.

12 Guerreiro, S. B., *et al*. 'Future heat-waves, droughts and floods in 571 European cities'. *Environmental. Research. Letters*. 13 034009, 2018.

13 Halliday, J. 'One in 10 new homes in England built on land with high flood risk'. *Guardian*, 19 February 2020.

14 Dunne, D. 'CO_2 emissions from wildfires have fallen over past 80 years, study finds'. www.carbonbrief.org, 17 April 2018.

15 Theobald, D. and Romme, W. 'Expansion of the US wildland–urban interface'. *Landscape and Urban Planning*, Vol. 83, Issue 4, 7 December 2007. pp. 340–54.

16 Hausfather, Z. 'Factcheck: How global warming has increased US wildfires'. www.carbonbrief.org, 9 August 2018.

17 Ghosh, P. 'Climate change boosted Australia bushfire risk by at least 30%'. *BBC News*, 4 March 2020.

18 Cox, L. 'Smoke cloud from Australian summer's bushfires three times larger than anything previously recorded'. *Guardian*, 2 November 2020.

19 Union of Concerned Scientists. 'The connection between climate change and wildfires'. www.ucsusa.org, 9 September 2011.

20 Goss, M., *et al*. 'Climate change is increasing the likelihood of extreme autumn wildfire conditions across California'. *Environmental Research Letters*, Vol. 15, No. 9, 20 August 2020.

21 Milman, O. 'Devastating 2020 Atlantic hurricane seasons breaks all records'. *Guardian,* 10 November 2020.

22 Henson, R. 2019. p. 194.

23 NOAA Geophysical Fluid Dynamics Laboratory. 'Global warming and hurricanes: an overview of current research results'. www.gfdl.noaa.gov, 29 March 2021.

24 Berardelli, J. 'How climate change is making hurricanes more dangerous'. Yale Climate Connections. 8 July 2019.

25 Knutson, T. R., *et al*. 'Climate change is probably increasing the intensity of tropical cyclones'. NOAA. www.climate.gov, 31 March 2021.

26 Emmanuel, K. 'Assessing the present and future probability of Hurricane Harvey's rainfall'. *Proceedings of the National Academy of Sciences of the United States of America*, Vol. 114, No. 48, 12681–4, 13 November 2017.

27 Nagchoudhary, S. and Paul, R. 'Cyclone Amphan loss estimated at $13 billion in India, may rise in Bangladesh'. *Reuters*, 23 May 2020; and Schwartz, M. 'Somalia's Strongest Tropical Cyclone Ever Recorded Could Drop 2 Years' Rain In 2 Days'. NPR, 22 November 2020.

28 https://www.worldweatherattribution.org/.

29 Rowlatt, J. 'Climate change: Siberian heatwave "clear evidence" of warming'. BBC News, 15 July 2020, https://www.bbc.co.uk/news/science-environment-53415297.

Chapter 10

1 Maslin, M. 2014. p. 103.
2 Henson, R. 2019. p. 125.
3 Brennan, P. 'Study: 2019 sees record loss of Greenland ice'. www.climate.nasa.gov, 20 August 2020.
4 Hirschi, J., Barnier, B., Böning, C., *et al.* 'The Atlantic meridional overturning circulation in high-resolution models'. *JGR Oceans,* Vol. 125, Issue 4, April 2020.
5 Maslin, M. 2014. p. 104.
6 Ritchie, P., Smith, G., Davis, K., *et al.* 'Shifts in national land use and food production in Great Britain after a climate tipping point'. *Nature Food*, Vol. 1, 76–83, 12 January 2020.
7 National Snow and Ice Data Center, 'Methane and Frozen Ground'. https://nsidc.org/cryosphere/frozenground/methane.html.
8 Herr, A., Osaka, S. and Stone, M. 'As the world warms, these Earth systems are changing. Could further warming make them spiral out of control?' www.grist.org.
9 Henson, R. 2019. p. 117.
10 Maslin, M. 2014. p. 109.
11 Alencar, A. and Esquivel Muelbert, A. 'The Amazon is now a net carbon producers, but there's still time to reverse the damage'. *Guardian,* 19 July 2021.
12 Carrington, A. 'Amazon rainforest now emitting more CO_2 than it absorbs'. *Guardian*, 14 July 2021.
13 Herr, A., Osaka, S. and Stone, M. 'As the world warms, these Earth systems are changing. Could further warming make them spiral out of control?' www.grist.org.

Chapter 11

1 Wynes and Nicholas. 2017.

2 Murtaugh and Schlax. 'Reproduction and the carbon legacies of individuals'. *Global Environmental Change*, 19 14–20, 2009.

3 BBC News. 'Stonehaven derailment: Report says climate change impact on railways "accelerating"'. www.bbc.co.uk/news, 10 September 2020.

Chapter 12

1 Ritchie, H. 'Sector by sector: where do global greenhouse gas emissions come from?'. www.ourworldindata.org, 18 September 2020.

2 EPA. 'Climate Change Indicators: Global Greenhouse Gas Emissions. April 2021.

3 Ritchie, H., and Roser, M. 'CO_2 and Greenhouse Gas Emissions". Published online at OurWorldInData.org. https://ourworldindata.org/co2-and-other-greenhouse-gas-emissions.

4 Ritchie, H. 'Who has contributed most to global CO_2 emissions?'. www.ourworldindata.org, 1 October 2019. Our World in Data based on Global Carbon Project; BP; Maddison; UNWPP.

5 Union of Concerned Scientists. 'Each Country's Share of CO_2 Emissions'. 12 August 2020. https://www.ucsusa.org/resources/each-countrys-share-co2-emissions.

6 Ritchie, H. 'How do CO_2 emissions compare when we adjust for trade?' 7 October 2019. https://ourworldindata.org/consumption-based-co2.

7 Our World In Data, 'Per Capita CO_2 emissions', based on Global Carbon Project; BP; Maddison; UNWPP.

8 United Nations, World Population Prospects 2019
9 Our World In Data, 'Per Capita CO_2 emissions', based on Global Carbon Project; BP; Maddison; UNWPP.

Chapter 13

1 IEA., 'World gross electricity production by source', 2019, IEA, Paris. https://www.iea.org/data-and-statistics/charts/world-gross-electricity-production-by-source-2019.
2 Ritchie, H. 'The death of UK coal in five charts'. www.ourworldindata.org, 28 January 2019.
3 US Energy Information Administration. 'Frequently asked questions: How much carbon dioxide is produced when different fuels are burned?'. www.eia.gov.
4 Alvarez, R., Pacala, S., Winebrake, J, Chameides, W. and Hamburg, S. 'Greater focus needed on methane leakage from natural gas infrastructure'. *PNAS*, Vol. 109, 17, 6435–440 , 24 April 2012.
5 Nunez, C. 'Fossil fuels, explained'. *National Geographic*, 2 April 2019.
6 IEA. 'Fuels and technology: Gas'. www.iea.org.
7 Vaughan, A. 'UK government rings death knell for the fracking industry'. 4 November 2019. https://www.newscientist.com/article/2222172-uk-government-rings-death-knell-for-the-fracking-industry/.
8 Hanlon, T. and Herz, N. 'Major oil companies take a pass on controversial lease sale in Arctic refuge'. www.npr.org, 6 January 2021.
9 IEA. *Carbon capture, utilization and storage*. www.iea.org, 2021.
10 IEA. *Net Zero by 2050: A roadmap for the global energy sector.* www.oea.org, 2021.

11 Coady, D., Parry, I., Le, N., Shang, B. *Global fossil fuel subsidies remain large: An update based on country-level estimates*, International Monetary Fund, 2 May 2019.

12 Lewis, S. L. *et al. Science*. 10.1126/science.aaz0388. 2019.

13 Tutton, M. 'The most effective way to tackle climate change? Plant 1 trillion trees'. www.edition.cnn.com, 17 April 2019.

14 Moomaw, W. R., *et al*. 'Wetlands in a changing climate: Science, policy and management'. *Wetlands*, 38, 183–205, 2018.

15 Joosten, H. 'The Global Peatland CO_2 Picture: Peatland status and drainage related emissions in all countries of the world'. Wetlands International, www.wetlands.org, 2010.

16 Hasegawa, T., Fujimoro, S., Havlík, P., *et al*. 'Risk of increased food insecurity under stringent global climate change mitigation policy'. *Nature Climate Change*, Vol. 8, 699–703, 30 July 2018.

17 Swain, F. 'The device that reverses CO_2 emissions'. 12 March 2021. https://www.bbc.com/future/article/20210310-the-trillion-dollar-plan-to-capture-co2.

18 Keutsch Group at Harvard. 'SCoPEx: Stratospheric controlled perturbation experiment'. www.keutschgroup.com.

Chapter 14

1 Royal Haskoning DHV. 'Norfolk Boreas Offshore Wind Farm Carbon Footprint Assessment'. 2020.

2 Li, H., Jiang, H., Dong, K., *et al*. 'A comparative analysis of the life cycle environment emissions from wind and coal power: Evidence from China'. *Journal of Cleaner Production*, Vol. 248, 119–2, 1 March 2020.

3 Gabbattiss, J. 'IEA: Wind and solar capacity will overtake both gas and coal globally by 2024'. www.carbonbrief.org, 10 November 2020.

4 BBC News. 'Renewables met 97% of Scotland's electricity demand in 2020'. www.bbc.co.uk/news, 25 March 2021.

5 Ambrose, J. 'UK electricity from renewables outpaces gas and coal power'. *Guardian*, 28 January 2021.

6 Multiple authors. 'In-depth: The UK should reach "net-zero" climate goal by 2050, says CCC'. Carbon Brief. 2 May 2019.

7 Roser,. M. 'Why did renewables become so cheap so fast? And what can we do to use this global opportunity for green growth?'. www.ourworldindata.org, 1 December 2020.

8 Lempriere, M. 'Solar PV costs fall 82% over the last decade, says IRENA'. www.solarportal.co.uk, 3 June 2020.

9 Henson, R. 2019. p. 433.

10 Office of Energy Efficiency & Renewable Energy. 'How much power is 1 gigawatt?'. www.energy.gov, 12 August 2019.

11 Davies, R. and Ambrose, J. 'Storm Bella helps Great Britain set new record for wind power generation'. *Guardian,* 28 December 2020.

12 Rincon, P. 'UK can be "Saudi Arabia of wind power" – PM'. www.bbc.co.uk/news, 24 September 2020.

13 Ambrose, J. 'Queen's property manager and Treasury to get windfarm windfall of nearly £9bn'. *Guardian*, 8 February 2021.

14 Ambrose, J. 'Why oil giants are swapping oil rigs for offshore windfarms'. *Guardian,* 10 February 2021.

15 Nield, D. 'Scotland is now generating so much wind energy it could power two Scotlands'. www.sciencealert.com, 17 July 2019.

16 SSE Renewables. 'Beatrice Offshore Wind Farm Limited'. www.sserenewables.com.

17 Buljan, A. 'Moray East Becomes Scotland's Largest OWF'. www.offshorewind.biz, 13 July 2021.

18 BBC News. 'Putin: Is he right about wind turbines and bird deaths?'. www.bbc.co.uk/news, 10 July 2019.

19 BBC News. 'Putin: Is he right about wind turbines and bird deaths?'

20 Byers, E. A., Coxon, G., Freer, J., *et al*. 'Drought and climate change impacts on cooling water shortages and electricity prices in Great Britain. *Nature Communications* 11, 2239, 7 May 2020, https://doi.org/10.1038/s41467-020-16012-2.

21 Berkeley Public Policy, The Goldman School. 'The US can reach 90 per cent clean electricity by 2035, dependably and without increasing consumer bills'. www.gspp.berkeley.edu, 9 June 2020.

22 https://www.mediamatters.org/fox-news/fox-news-and-fox-business-falsely-blamed-renewable-energy-texas-blackouts-128-times-over.

23 https://www.texastribune.org/2021/02/16/texas-wind-turbines-frozen/.

24 Evans, S. 'In-depth: How a smart flexible grid could save the UK £40bn'. www.carbonbrief.org, 25 July 2017.

25 Verger, R. 'Tesla actually built the world's biggest battery. Here's how it works'. www.popsci.com, 2 December 2017.

26 Blain, L. 'Australia plans world's biggest battery (again), at 1.2 gigawatts'. www.newatlas.com, 4 February 2021.

27 Ryu, A. and Meshkati, N. 'Onagawa: The Japanese nuclear power plant that didn't melt down on 3/11'. www.thebulletin.org, 10 March 2014.

28 Frangoul, A. 'From powerful tidal turbines to huge wave machines, Scotland is becoming a hub for marine energy'. www.cnbc.com, 25 May 2021.

29 Burgen, S. '"A role model": how Seville is turning leftover oranges into electricity'. *Guardian*, 23 February 2021.

30 Science Daily. 'New wearable device turns the body into a battery'. www.sciencedaily.com, 10 February 2021.

Chapter 16

1 Henson, R. 2019. p. 477.

2 Pollard, T. 'Number of cars on UK roads surpasses 40 million for first time'. www.carmagazine.co.uk, 21 April 2021.

3 ONS population estimates. www.ons.gov.uk, 25 June 2021.

4 WRAP. 'Net zero: Why resource efficiency holds the answers'. www.wrap.org, 2021.

5 Phys.org. 'Five things to know about VW's "dieselgate" scandal'. www.phys.org, 18 June 2018.

6 BBC News. 'Ex-Audi boss stands trial over "dieselgate" scandal in Germany'. www.bbc.co.uk/news, 30 September 2020.

7 Milman, O. 'Massachusetts city to post climate change warning stickers at gas stations'. *Guardian*, 25 December 2020.

8 Laville, S. 'Ban SUV adverts to meet UK climate goals, report urges'. *Guardian* 3 August 2020.

9 Cozzi, L. and Petropoulos, A. 'Growing preference for SUVs challenges emissions reductions in passenger car market'. www.iea.org, 15 October 2019.

10 Cozzi, L. and Petropoulos, A. 2019.

11 BBC News. 'Rise of SUVs "makes mockery" of electric car push'. www.bbc.co.uk/news, 9 December 2019.

12 Pidd, H. 'School pupils issue fake parking tickets to tackle pollution'. *Guardian,* 13 February 2019.

13 WWF, 'Le trop plein de SUV dans la publicité'. March 2021.

14 Harrabin, R. 'Adverts for large polluting cars "should be banned"'. www.bbc.co.uk/news, 3 August 2020.

15 Chen, D. and Kockelman, K. 'Carsharing's life-cycle impacts on energy use and greenhouse emissions'. *Transportation Research Part D: Transport and Environment*, Vol. 47, August 2016. pp. 276–84.

16 Brand, C. 'Blog: How your legs can reduce your carbon footprint'. www.ukerc.ac.uk, 4 February 2021.

17 Partridge, J. 'Electric cars "will be cheaper to produce than fossil fuel vehicles by 2027"'. *Guardian,* 9 May 2021.

18 Consultancy.eu. 'Europe's electric vehicles fleet to reach 40 million by 2030'. www.consultancy.eu, 25 February 2021.

19 Neate, R. 'Ford plans for all cars sold in Europe to be electric by 2030'. *Guardian*, 17 February 2021.

20 Reuters. 'Electric cars rise to record 54% market share in Norway'. *Guardian*, 5 January 2021.

21 Bonnici, D. 'How A-ha made Norway take on EVs'. *Which Car?*, 11 January 2021.

22 @AukeHoekstra (Auke Hoekstra). Thread: 'New "study" claims it takes 48k miles for electric vehicles to emit less CO_2 than gasoline cars. But it's just a misleading brochure . . .' 27 November 2020, https://twitter.com/AukeHoekstra/status/1332464525602410498.'

23 Transport & Environment. 'How clean are electric cars?'. "http://www.transportandenvironment.org" www.transportandenvironment.org

24 Office of Energy Efficiency & Renewable Energy. 'Battery-electric vehicles have lower scheduled maintenance costs

than other light-duty vehicles'. www.energy.gov, 14 June 2021.

25 Grundy, A. 'Tesco EV charging rollout hits new milestone with 500,000 charges on network'. www.current-news.co.uk, 9 April 2021.

26 Bannon, E. 'Postcode lottery for electric car charging must be fixed – NGO'. www.transportenvironment.org, 19 May 2021.

Chapter 17

1 Kommenda, N. 'How your flight emits as much CO_2 as many people do in a year'. *Guardian,* 19 July 2019.

2 Timperley, J. 'Should we give up flying for the sake of the climate?' www.bbc.com, 19 February 2020.

3 Lenzen, M., Sun, Y., Faturay, F., Ting, Y., Geschke, A. and Malik, A. 'The carbon footprint of global tourism'. *Nature Climate Change*, Vol. 8, 522–28, 7 May 2018.

4 Bannon, E. 'Biofuels policies to massively increase deforestation by 2030 – study'. www.transportenvironment.com. 19 March 2020.

5 Hotten, R. 'Ryanair rapped over low emissions claims'. www.bbc.co.uk/news, 5 February 2020.

6 International Air Transport Association. 'IATA forecast predicts 8.2 billion air travellers in 2037'. www.iata.org, 24 October 2018.

7 Timperley, J. 2020.

8 Barrett, T. 'Exclusive report: high time airlines paid tax on fuel?'. www.airqualitynews.com, 10 May 2019.

9 De Clercq, G. 'France wants EU to seek end to jet fuel tax exemption to curb emissions'. *Reuters*, 3 June 2019.

10 Harrabin, R. 'A few frequent flyers "dominate air travel"'. www.bbc.co.uk/news, 31 March 2021.

11 Coffey, H. 'Private jet flights from UK and France emit more CO_2 than 20 other European countries'. *Independent*, 27 May, 2021.

12 Bannon, E. 'Private jets: can the super-rich supercharge zero-emission aviation?'. www.transportenvironment.org, 27 May 2021.

13 Gössling, S. 'Celebrities, air travel and social norms'. *Annals of Tourism Research*, Vol. 79, 102775, November 2019.

14 Adams, C. 'More British people flew abroad last year than any other nationality, new data reveals'. *Independent*, 31 July 2019.

15 EEA. 'Motorised transport: train, plane, road or boat – which is greenest?' www.eea.europa.eu. 24 March 2021.

16 Reality Check. 'Climate change: Should you fly, drive or take the train?' *BBC News*, 24 August 2019.

17 Reuters. 'French lawmakers approve a ban on short domestic flights'. www.reuters.com, 11 April 2021.

18 Hughes, A. 'Young adults most concerned about green travel, survey says'. *Independent*, 15 February 2021.

19 The Man in Seat 61. www.seat61.com.

20 'Rogelj, J., Geden, O., Cowie, A., and Reisinger, A. 'Net-zero emissions targets are vague: three ways to fix'. *Nature* 591, 365–8. 2021.

21 Cairns, S., Patrick, J. and Newson, C. 2021.

22 Lund, T. 'Sweden's air travel drops in year when "flight shaming" took off'. www.reuters.com, 10 January 2020.

23 Buchs, M. and Mattioli, G. 'Trends in air travel inequality in the UK: From the few to the many?' www.creds.ac.uk, 7 July 2021; and Transport & Environment. 'Flying and climate change'. www.transportenvironment.org.

Chapter 19

1 Piddington, J., Nicol, S., Garrett, H. and Custard, M. *The Housing Stock of the United Kingdom*. BRE Trust, February 2020.

2 Committee on Climate Change. *Next steps for UK heat policy*. Committee on Climate Change, October 2016.

3 Committee on Climate Change. *UK housing: Fit for the future?* Committee on Climate Change, February 2019.

4 Committee on Climate Change, 2019.

5 Confederation of British Industry (CBI). 'Net-zero: the road to low-carbon heat'. www.cbi.org.uk, 22 July 2020.

6 Harkin, D. 'Using the past to inspire the future'. www.historicenvironment.scot, 26 October 2018.

7 International Energy Agency. 'The future of cooling'. www.iea.org, May 2018.

8 International Energy Agency. 'Air conditioning use emerges as one of the key drivers of global electricity-demand growth'. www.iea.org, 15 May 2018.

9 US Department of Energy. 'LED lighting'. www.energy.gov.

10 Khosla, R., Shrish Kamat, A. and Narayanamurti, V. 'Guest post: How energy-efficient LED bulbs lit up India in just five years'. www.carbonbrief.org, 31 March 2020.

11 Blunden, M. 'Map reveals London's best and worst neighbourhoods for energy efficiency'. *Evening Standard*, 11 September 2020.

12 Climate Change Committee. *Reducing UK emissions: Progress Report to Parliament*. Climate Change Committee, June 2020.

13 Antonelli, L. 'How Brussels went passive'. www.passivehouseplus.ie, 26 October 2016.

14 Perrott, R. 'How many cats would it take to heat a Passive House?' www.c60design.co.uk.

15 Lowe, R. and Oreszczyn, T. *Building decarbonisation transition pathways: initial reflections*. CREDS Policy brief 013. Centre for Research into Energy Demand Solutions, 2020.
16 Vaughan, A. 'Fix the planet' (email newsletter), www.newscientist.com.
17 Lowe, R. and Oreszczyn, T. 2020.

Chapter 20

1 Fecht, S. 'Wine regions could shrink dramatically with climate change unless growers swap varieties'. www.news.climate.colombia.edu, 27 January 2020.
2 Poore, J. and Nemecek, T. 'Reducing food's environmental impacts through producers and consumers'. *Science*, Vol. 360, Issue 6392, 1 June 2018. pp. 987–92.
3 Department of Economic and Social Affairs, United Nations. *World Population Prospects 2019*. www.population.un.org.
4 Smith, P. ' Malthus is still wrong: we can feed a world of 9–10 billion, but only by reducing food demand'. Proceedings of the Nutrition Society, 74, 187–190, (2015), doi:10.1017/S0029665114001517.
5 Food and Agriculture Organization of the United Nations. *Food wastage footprint: Impacts on natural resources*. FOA, 2013.
6 Oakes, K. 'How cutting your food waste can help the climate'. www.bbc.com, 26 February 2020.
7 WRAP. 'Food waste falls by 7% per person in three years'. www.wrap.org.uk, 24 January 2020.
8 Nair, P. 'The country where unwanted food is selling out'. www.bbc.com, 24 January 2017.
9 Oakes, K. 2020.

10 Porter, S., Reay, D., Bomberg, E. and Higgins, P. 'Available food losses and associate production-phase greenhouse gas emissions arising from application of cosmetic standards to fresh fruit and vegetables in Europe and the UK'. *Journal of Cleaner Production*, Vol. 201, 10 November 2018. pp. 869–78.

11 Willet, W., Rockström, J., Loken, B., *et al*. 'Food in the Anthropoecene: the EAT-Lancet Commission on healthy diets from sustainable food systems'. *The Lancet Commissions*, Vol. 393, Issue 10170, 2 February 2019. pp. 447–92.

12 Carrington, D. 'No-kill, lab-grown meat to go on sale for first time'. *Guardian,* 2 December 2020.

13 Myhre, G., D. Shindell, Bréon, F.-M., Collins,W., Fuglestvedt, J., Huang, J., Koch, D., Lamarque, J.-F., Lee, D., Mendoza, B., Nakajima, T., Robock, A., Stephens, G., Takemura, T. and Zhang, H. 'Anthropogenic and Natural Radiative Forcing', *Climate Change 2013: The Physical Science Basis. Contribution of Working Group I to the Fifth Assessment Report of the Intergovernmental Panel on Climate Change* (Stocker, T. F., Qin, D., Plattner, G.-K., Tignor, M., Allen, S.K., Boschung, J., Nauels, A., Xia, Y., Bex, V. and Midgley, P.M., eds.). Cambridge University Press, 2014.

14 Boucher, D. 'Movie review: There's a vast Cowspiracy about climate change'. Union of Concerned Scientists, www.blog.uscusa.org, 10 June 2016.

15 Twine, R. 'Emissions from Animal Agriculture – 16.5% Is the New Minimum Figure'. *Sustainability*. *13*(11), 6276. 2021.

16 Brown, D. 'Five ways UK farmers are tackling climate change'. www.bbc.co.uk/news, 9 September 2019.

17 Rajão, R., *et al*. 'The rotten apples of Brazil's agribusiness' *Science*'. Vol. 369, Issue 6501, pp. 246–8. 17 July 2020.

18 Milman, O. 'Feeding cows seaweed could cut their methane emissions by 82%, scientists say'. *Guardian*, 18 March 2021.

19 Veeramani, A., Dias, G. and Kirkpatrick, S. 'Carbon footprint of dietary patterns in Ontario, Canada: A case study based on actual food consumption'. *Journal of Cleaner Production*, Vol. 162, 20 September 2017. pp. 1398–1406.

20 Tilman, D. and Clark, M. 'Global diets link environmental sustainability and human health'. *Nature*, Vol. 515, 12 November 2014. pp. 518–22.

21 Springmann, M. *et al*. 'Analysis and valuation of the health and climate change co-benefits of dietary change' *PNAS*. 113 (15). 4146–51. 12 April 2016.

22 Maslin, M. and Nab, C. 'Coffee: here's the carbon cost of your daily cup – and how to make it climate-friendly'. *The Conversation*. 4 January 2021.

23 Poore, J. and Nemecek, T. 'Reducing food's environmental impacts through producers and consumers'. *Science*, Vol. 360, Issue 6392, 1 June 2018. pp. 987–92.

24 McGivney, A. 'Almonds are out. Dairy is a disaster. So what milk should we drink?' *Guardian*, 29 January 2020.

25 Cohen, D. 'Unilever: Breakthrough as food industry giant introduces carbon footprint labels on food'. *Independent*, 15 July 2021.

26 Harvey, F. 'Outrage and delight as France ditches reliance on meat in climate bill'. *Guardian*, 29 May 2021.

Chapter 21

1 Ritchie, H. 'FAQs on plastics'. www.ourworldindata.org, 2 September 2018.

2 Ritchie, H. and Roser, M. 'Plastic pollution'. www.ourworldindata.org, September 2018.

3 Elhacham, E., Ben-Uri, L., Grozovski, J., Bar, Y. and Milo, R. 'Global human-made mass exceeds all living biomass'. *Nature*, Vol. 588, 9 December 2020.

4 Lebreton, L., Slat, B., Ferrari, F., *et al.* 'Evidence that the Great Pacific Garbage Patch is rapidly accumulating plastic'. *Scientific Reports* 8, 4666 (2018). https://doi.org/10.1038/s41598-018-22939-w.

5 Carrington, D. 'Microplastics revealed in the placentas of unborn babies'. *Guardian*, 22 December 2020.

6 McDonalds. 'McDonald's pledges to remove non-sustainable hard plastic from its iconic Happy Meal'. www.mcdonalds.com/gb. 17 March 2020.

7 *Dispatches*. 'The Dirty Truth About Your Rubbish'. Channel 4. Producer/Director: Andrew Pugh. 8 March 2021.

8 Stoufer, L. 'Plastics packaging: today and tomorrow'. Report presented at the 1963 Society of the Plastics Industry, Inc. Annual National Plastics Conference. Sheraton-Chicago Hotel, Chicago, Illinois. 19–21 November 1963.

9 McKay, D. 'Fossil fuel industry sees the future in hard-to-recycle plastic'. *The Conversation*, 10 October 2019.

10 Wynes, S. and Nicholas, K., 2017.

11 Environment Agency. 'An updated lifecycle assessment study for disposable and reusable nappies'. Science Report SC010018/SR2. October 2008.

12 Sohn, J., Nielsen, K., Birkved, M., Joanes, T. and Gwozdz, W. 'The environmental impacts of clothing: Evidence from United States and three European countries'. *Sustainable Production and Consumption*, Vol. 27, July 2021. pp. 2153–64.

13 Sohn, J., *et al.* July 2021.

14 Berners-Lee, M. 'How Bad Are Bananas: The Carbon Footprint Of Everything' Profile Books. Revised 2020 Edition.

15 Savage, M. 'How can we make washing machines last?'. www.bbc.co.uk/news, 3 March 2021.

16 Jeswani, H. K., and Azapagic, A. 'Is e-reading environmentally more sustainable than conventional reading?' *Clean Technologies and Environmental Policy*, Vol. 17, Issue 3, March 2014. pp. 803–9.

17 IEA. 'Data Centres and Data Transmission Networks'. IEA, Paris. 2020. https://www.iea.org/reports/data-centres-and-data-transmission-networks.

18 Cambridge Centre for Alternative Finance. 'Cambridge Bitcoin Electricity Consumption Index'. Online at https://cbeci.org/cbeci/comparisons.

19 @BitcoinMagazine (Bitcoin Magazine). Thread: 'While Elon Musk claimed that Tesla is "concerned about rapidly increasing use of fossil fuels for Bitcoin mining and transactions", the mining industry appears to be growing in its use of renewable energy sources instead'. 13 May 2021, https://twitter.com/BitcoinMagazine/status/1392899567641845760?s=20.

20 Ecosia. 'Ecosia financial reports'. www.blog.ecosia.org , June 2021.

21 Google. *Google Environmental Report 2019*. www.sustainability.google, 2019.

22 Hern, A. 'Facebook says it has reached net zero emissions'. *Guardian*, 16 April 2021.

Chapter 22

1 Watts, N., Amann, M., Arnell, N., *et al*. 'The 2019 report of the Lancet Countdown on health and climate change:

ensuring that the health of a child born today is not defined by a changing climate'. *The Lancet*, Vol. 394, Issue 10211, 16 November 2019. pp. 1836–78.

Chapter 23

1 Dosio, A., *et al*. 'Extreme heat waves under 1.5 °C and 2 °C global warming'. *Environmental Research Letters* 13 054006, 2018.

2 IPCC. 'Special report: Global warming of 1.5°C'. www.ipcc.ch.

3 UN Environment Programme and DTU Partnership. *UNEP Emissions Gap Report 2020*. www.unep.org, 9 December 2020.

4 Gerretsen, I. 'Germany raises ambition to net zero by 2045 after landmark court ruling'. www.climatechangewnews.com, 5 May 2021.

5 Jaeger, J., McLaughlin, K., Neuberger, J. and Dellesky, C. 'Does Biden's American jobs plan stack up on climate and jobs?'. www.wri.org, 1 April 2021.

6 Harvey, F. 'China "must shut 600 coal-fired plants" to hit climate target'. *Guardian*, 15 April 2021.

7 Pike, L. 'Why China is still clinging to coal'. www.vox.com, 6 April 2021.

8 UN Environment Programme and DTU Partnership, 2020.

9 Department for Business, Energy & Industrial Strategy. 'Press release: UK sets ambitious new climate target ahead of summit'. www.gov.uk, 3 December 2020.

10 Ioualalen, R. 'Spain becomes latest country to ban new oil and gas exploration and production'. www.priceofoil.org, 14 May 2021.

11 Strzyżyńska, W. 'Sámi reindeer herders file lawsuit against Norway windfarm'. *Guardian*, 18 January 2021.

12 Hoffower, H. and Hartmans, A. 'Bill and Melinda Gates are ending their 27-year marriage. Here's how the Microsoft co-founder spends his $129 billion fortune, from a luxury-car collection to incredible real estate'. www.businessinsider.com, 3 May 2021.

13 McCollum, D. L., *et al*. 'Energy investment needs for fulfilling the Paris Agreement and achieving the sustainable development goals'. *Nature Energy* 3, 589–99. 2018.

14 Ameli, N., Dessens, O., Winning, M., *et al*. 'Higher cost of finance exacerbates a climate investment trap in developing economies. *Nature Communications*, 12: 4046. 2021.

15 Podesta, J., Goldfuss, C., Higgins, T., *et al*. 'State Fact Sheet: A 100 per cent clean future'. www.americanprogress.org, 16 October 2019.

16 Office of Governor Gavin Newsom. 'Governor Newsom announces California will phase out gasoline-powered cars and drastically reduce demand for fossil fuel in California's fight against climate change'. www.gov.ca.gov, 23 September 2020.

17 Coalition for Urban Transitions, www.urbantransitions.global.

18 Robertson, D. 'Inside Copenhagen's race to be first carbon-neutral city'. *Guardian*, 11 October 2019.

19 Halais, F. 'Cities race to slow climate change – and improve life for all'. www.wired.com, 1 January 2020.

20 BBC News. 'Glasgow and Edinburgh fight to become the UK's first "net-zero" city'. www.bbc.co.uk/news, 15 May 2019.

21 CDP. 'Manchester: How the UK's "City-Region of Change" is setting the bar for climate action'. www.cdp.net.

22 City of Sydney. 'Net zero by 2035: our bold new plan'. 2021.

https://news.cityofsydney.nsw.gov.au/articles/net-zero-by-2035-city-sydney-bold-new-plan.

23 Energy & Climate Intelligence Unit. 'Report: Fifth of world's largest companies now have net zero target'. www.eciu.net, 23 March 2021.

24 Ingrams, S. 'IKEA will now buy back your old furniture in new sustainability scheme'. *Which?*, 5 May 2021.

25 Nespresso. 'Our climate commitment'. www.sustainability.nespresso.com.

26 Climate Action Tracker 'Global update: Climate Summit Momentum. www.climateactiontracker.org, 4 May 2021.

Chapter 24

1 Centre for Climate Change and Social Transformations (CAST). *Survey infographic: UK perceptions of climate change & lifestyle changes*. Cardiff University, 2021.

2 Marshall, G. *Don't Even Think About It: Why Our Brains Are Wired to Ignore Climate Change*. Bloomsbury, 2015. p. 137.

3 Marshall, 2015. p. 64.

4 Stoknes, P. E. *What We Think About When We Try Not To Think About Global Warming*. Chelsea Green Publishing, 2015.

5 December 2020 figures, https://climatecommunication.yale.edu/about/projects/global-warmings-six-americas/.

6 https://rare.org/blog/qa-with-climate-scientist-katharine-hayhoe/.

Chapter 26

1 Franta, B. 'Early oil industry knowledge of CO_2 and global warming'. *Nature Climate Change*, Vol. 8, 19 November 2018. pp. 1024–5.

2 Banerjee, N., Song, L. and Hasemyer, D. 'Exxon's own research confirmed fossil fuels' role in global warming decades ago'. www.insideclimatenews.org, 16 September 2015.
3 Franta, B. 'Shell and Exxon's secret 1980s climate change warnings'. *Guardian*, 19 September 2018.
4 DeSmog. 'Global Climate Coalition'. www.desmog.com.
5 Shabecoff, P. 'Global warming has begun, expert tells senate'. *New York Times*, 24 June 1988.
6 Vidal, J. 'Revealed: how oil giant influenced Bush'. *Guardian*, 8 June 2005.
7 Climate Files. '1991 Information Council on the Environment Climate Denial Ad Campaign'. www.climatefiles.com.
8 Hasemyer, D. and Cushman Jr., J. H. 'Exxon Sowed Doubt About Climate Science for Decades by Stressing Uncertainty', *Inside Climate News*, www. insideclimatenews.org, 22 October 2015.
9 Oreskes, N. and Conway, E. *Merchants of Doubt: How a handful of scientists obscured the truth on issues from tobacco smoke to global warming*. Bloomsbury, 2012.
10 Shearer, C. *Kivalina: A Climate Change Story*. Haymarket Books, 2011.
11 Greenpeace., 'Exxon's Climate Denial History: A Timeline'. Accessed 23 August 2021, https://www.greenpeace.org/usa/ending-the-climate-crisis/exxon-and-the-oil-industry-knew-about-climate-change/exxons-climate-denial-history-a-timeline/.
12 Adam, D. 'Exxon to cut funding to climate change denial groups'. *Guardian*, 28 May 2008.
13 Climate One. 'My Climate Story: Ben Santer'. www.climateone.org, 17 September 2019.

14 Grandia, K. 'The 30,000 global warming petition is easily debunked propaganda'. www.huffpost.com, 22 August 2009.

15 DeSmog., 'Oregon Petition'. www.desmog.com.

16 BBC News. 'BBC climate change interview breached broadcasting standards'. www.bbc.co.uk/news, 9 April 2018.; https://www.desmog.com/nigel-lawson/.

17 Supran, G., and Oreskes, N. 'Assessing ExxonMobil's climate change communications (1977–2014)'. *Environmental Research Letters*, Vol. 12, No. 8, 23 August 2017.

18 Sutherland, J. 'They call it pollution. We call it life'. www.npr.org, 23 May 2006.

19 Skeptical Science. 'Clearing up misconceptions regarding "hide the decline"'. www.skepticalscience.com.

20 Yeo, S. 'Why the Climategate hack was more than an attack on science'. www.desmog.com, 15 November 2019.

21 McKie, R. 'Climategate 10 years on: what lessons have we learned?' *Observer*, 9 November 2019.

22 McEvers, K. 'Saudi Arabia tries to stall global emissions limits'. www.npr.org, 10 December 2009.

23 Influence Map. 'Big oil's real agenda on climate change', www.influencemap.org, March 2019. https://influencemap.org/report/How-Big-Oil-Continues-to-Oppose-the-Paris-Agreement-38212275958aa21196dae3b76220bddc.

24 Carrington, D. 'How to spot the difference between a real climate policy and greenwashing guff'. *Guardian*, 6 May 2021.

25 Mann, M. *The New Climate War: The Fight to Take Back Our Planet*. Scribe, 2021.

26 Gearino, D. 'Inside clean energy: 6 Things Michael Moore's *Planet of the Humans* gets wrong'. www.insideclimatenews, 30 April 2020.

27 Joshi, K. 'Planet of the humans: A reheated mess of lazy old myths'. www.ketanjoshi.co, 24 April 2020.

28 Chewpreecha, U. and Summerton, P. *Economic impact of the Sixth Carbon Budget*. Climate Change Committee, Cambridge Econometrics, December 2020.

29 Office for Budget Responsibility. *Fiscal Risks Report*. www.obr.uk, July 2021.

30 Client Earth. 'What is greenwashing? An interview with Sophie Marjanac'. www.clientearth.org, 4 November 2020.

31 Influence Map. 'Big oil's real agenda on climate change'. www.influencemap.org, March 2019. https://influencemap.org/report/How-Big-Oil-Continues-to-Oppose-the-Paris-Agreement-38212275958aa21196dae3b76220bddc.

32 Carrington, D. '"A great deception"; oil giants taken to task over "greenwash" ads'. *Guardian*, 19 April 2021.

33 Friedlingstein, P., O'Sullivan, M., Jones, M. W., Andrew, R. M., Hauck, J., Olsen, A., Peters, G. P., Peters, W., Pongratz, J., Sitch, S., Le Quéré, C., Canadell, J. G., Ciais, P., Jackson, R. B., Alin, S., Aragão, L. E. O. C., Arneth, A., Arora, V., Bates, N. R., Becker, M., Benoit-Cattin, A., Bittig, H. C., Bopp, L., Bultan, S., Chandra, N., Chevallier, F., Chini, L. P., Evans, W., Florentie, L., Forster, P. M., Gasser, T., Gehlen, M., Gilfillan, D., Gkritzalis, T., Gregor, L., Gruber, N., Harris, I., Hartung, K., Haverd, V., Houghton, R. A., Ilyina, T., Jain, A. K., Joetzjer, E., Kadono, K., Kato, E., Kitidis, V., Korsbakken, J. I., Landschützer, P., Lefèvre, N., Lenton, A., Lienert, S., Liu, Z., Lombardozzi, D., Marland, G., Metzl, N., Munro, D. R., Nabel, J. E. M. S., Nakaoka, S.-I., Niwa, Y., O'Brien, K., Ono, T., Palmer, P. I., Pierrot, D., Poulter, B., Resplandy, L., Robertson, E., Rödenbeck, C., Schwinger, J., Séférian, R., Skjelvan, I.,

Smith, A. J. P., Sutton, A. J., Tanhua, T., Tans, P. P., Tian, H., Tilbrook, B., van der Werf, G., Vuichard, N., Walker, A. P., Wanninkhof, R., Watson, A. J., Willis, D., Wiltshire, A. J., Yuan, W., Yue, X., and Zaehle, S. 'Global Carbon Budget 2020'. Earth Syst. Sci. Data, 12, 3269–3340, https://doi.org/10.5194/essd-12-3269-2020, 2020.

34 Hausfather, Z. 'Analysis: When might the world exceed 1.5°C and 2°C of global warming?' www.carbonbrief.org, 4 December 2020.

35 Bousso, R. 'Shell to write down assets again, taking cuts to more than $22 billion'. www.reuters.com, 21 December 2020.

36 Client Earth. 'The Greenwashing Files'. www.clientearth.org.

37 Joshi, K. 'A major test for Shell's massive multi-purpose greenwashing juggernaut'. www.medium.com, 29 April 2021.

38 Boffey, D. 'Court orders Royal Dutch Shell to cut carbon emissions by 45% by 2030'. *Guardian*, 26 May 2021.

39 Reguly, E. 'A tale of transformation: the Danish company that went from black to green energy'. www.corporateknights.com, 16 April 2019.

40 Ambrose, J. 'BP market value at 26-year low amid investor jitters'. *Guardian*, 21 October 2020.

Chapter 27

1 Harvey, F. 'Campaign seeks 1bn people to save climate – one small step at a time', *Guardian*, 10 October 2020; and Count Us In, www.count-us-in.org.

2 Doyle, J. 'Where has all the oil gone? BP branding and the discursive elimination of climate change risk'. *Culture, Environment and Eco-Politics*, eds. Heffernan, N. and Wragg, D. Cambridge Scholars, January 2011, pp. 200–25.

3 UN Environment Programme and DTU Partnership, 2020.

4 Guenther, G. 'Who is the *we* in "We are causing climate change"?' www.slate.com, 10 October 2018.

5 Bollinger, B. and Gillingham, K. 'Peer effects in the diffusion of solar photovoltaic panels'. *Marketing Science*, Vol. 31, No. 6, 20 September 2012. pp. 873–1025.

6 Rainforest Action Network. *Banking on Climate Chaos: Fossil Fuel Finance Report 2021*. www.ran.org.

7 Switch It. www.switchit/money.

8 @KHayhoe (Prof. Katharine Hayhoe). Tweet: 'A thermometer isn't liberal or conservative; it doesn't give you a different number depending on whether you vote left or right.' 3 December 2016, https://twitter.com/KHayhoe/status/804856223661223936?s=20.

9 Steentjes, K., Poortinga, W., Demski, C., and Whitmarsh, L. *UK perceptions of climate change and lifestyle changes.* CAST Briefing Paper 08. 2021.

10 BEIS (UK Department of Business, Energy and Industrial Strategy). *Public Attitudes Tracker: Wave 29*. www.gov.uk, 9 May 2019.

11 Centre for Towns. *More United Than You'd Think: Public Opinion on the Environment in Towns and Cities in the UK*. www.centrefortowns.org, 9 December 2020.

12 Marris, E. 'Why young climate activists have captured the world's attention'. www.nature.com, 18 September 2019.

13 Lawson, D., Stevenson, K., Peterson, N., *et al*. 'Children can foster climate change concern among their parents'. *Nature Climate Change*, Vol. 9, 6 May 2019. pp. 457–62.

14 Murray, J. '"It's awakened me": UK climate assembly participants hail a life-changing event'. *Guardian*, 31 December 2020.

15 CDP. 'From climate emergencies to climate action'. www.cdp.net.

16 Climate Emergency Declaration. 'Climate emergency declarations in 2,010 jurisdictions and local governments cover 1 billion citizens'. www.climateemergencydeclaration.org, 1 August 2021.

17 Community Led Homes. 'Goodwin Development Trust'. www.communityledhomes.org.uk.

18 Taylor, M. 'How grassroots schemes across UK are tackling climate crisis'. *Guardian*, 10 March 2021.

19 Webb, J., Stone, L., Murphy, L. and Hunter, J. *The Climate Commons: Home communities can thrive in a climate changing world*. Institute for Public Policy Research, March 2021.

20 Cairns, I., Hannon, M., *et al*. 'Financing community energy case studies: Green Energy Mull'. UK Energy Research Centre, June 2020.

21 Webb, J., *et al*. March 2021.

22 Heglar, M. 'We can't tackle climate change without you'. www.wired.com, 4 January 2020.